THE SEEDS OF DICOTYLEDONS
VOLUME 2

THE SEEDS OF DICOTYLEDONS

E. J. H. CORNER, F.R.S.

EMERITUS PROFESSOR OF TROPICAL BOTANY, UNIVERSITY OF CAMBRIDGE

VOLUME 2
ILLUSTRATIONS

CAMBRIDGE UNIVERSITY PRESS
CAMBRIDGE
LONDON · NEW YORK · MELBOURNE

CAMBRIDGE UNIVERSITY PRESS
Cambridge, New York, Melbourne, Madrid, Cape Town, Singapore, São Paulo, Delhi

Cambridge University Press
The Edinburgh Building, Cambridge CB2 8RU, UK

Published in the United States of America by Cambridge University Press, New York

www.cambridge.org
Information on this title: www.cambridge.org/9780521116039

First published 1976
This digitally printed version 2009

A catalogue record for this publication is available from the British Library

Library of Congress Cataloguing in Publication data
Corner, Edred John Henry.
The seeds of dicotyledons.
Includes bibliographical references and index.
1. Dicotyledons. 2. Seeds. I. Title.
QK495.A12C67   583′.04′16   74–14434

ISBN 978-0-521-20687-7 hardback
ISBN 978-0-521-11603-9 paperback

CONTENTS

VOLUME 2

The contents of this volume consist of the Figures described in Volume 1

VOLUME 1

Preface   *page* vii

PART ONE

1 **Material and method**   1

2 **Seed-form**   3
The bitegmic anatropous seed, 4; campylotropous seeds, 4; obcampylotropous seeds, 4; hilar seeds, 4; pre-raphe seeds, 4; orthotropous seeds, 5; the dorsal raphe, 5; perichalazal seeds, 5; pachychalazal seeds, 5; alate seeds, 6; pleurogrammatic seeds, 7

3 **Seed-coats**   8
Testa and tegmen, 8; multiplicative integuments and overgrown seeds, 8; factors in the formation of seeds, 9; description of the seed-coats, 9; exotestal seeds, 10; mesotestal seeds, 11; endotestal seeds; crystal-cells and raphid-cells, 11; exotegmic seeds with a palisade, 13; exotegmic seeds with fibres, 14; exotegmic tracheidal seeds, 17; endotegmic seeds, 18; undifferentiated seed-coats, 18; cell-form, 19; aerenchyma and stomata, 19; the vascular supply of the seed, 20; hairs, 22; chalaza, 22; aril, 23; sarcotesta, 24

4 **Criticism of the arrangement of dicotyledonous families into orders**   25
Magnoliales–Ranales, 25; Ranunculales, 27; Theales–Guttiferae–Dilleniales, 28; Bixales–Violales, 30; Rutales–Sapindales, 30; Celastrales, 32; Capparidales–Cruciales–Rhoeadales, 32; Euphorbiales–Malvales–Thymelaeales–Tiliales, 33; Geraniales–Malpighiales–Polygalales, 35; Hamamelidales–Rosales–Leguminosae, 35; Rhamnales–Proteales, 36; Lythrales–Myrtales, 37; Cucurbitales–Passiflorales, 38; Aristolochiales–Nepenthales–Rafflesiales–Sarraceniales, 40; Araliales–Cornales–Caprifoliaceae, 40; Centrospermae, 41; Tamaricales, 42; Amentiferae, 42; Piperales, 43; summary of positive contributions, 44; classification of bitegmic dicotyledonous seeds, 45

5 **Seed-evolution**   48
The primitive dicotyledonous seed, 48; the evolution of the dicotyledonous seed, 48; unitegmic seeds: the origin of Sympetalae, 49; Convolvulaceae, 50; arillate and sarcotestal seeds, 52; seed-size, 52; a Myristicalean start, 55; seed-progress, 57; transference of function, 58; the origin of the angiosperm seed, 58; neoteny again, 59

PART TWO

Descriptions of seeds by families   65
Acanthaceae, 65; Aceraceae, 65; Actinidiaceae, 65; Adoxaceae, 66; Aextoxicaceae, 66; Aizoaceae, 66; Akaniaceae, 67; Alangiaceae, 67; Amaranthaceae, 67; Anacardiaceae, 67; Ancistrocladaceae, 68; Annonaceae, 68; Apocynaceae, 70; Aquifoliaceae, 73; Araliaceae, 73; Aristolochiaceae, 73; Asclepiadaceae, 74
Balanitaceae, 75; Balsaminaceae, 75; Basellaceae, 75; Begoniaceae, 75; Berberidaceae, 75; Betulaceae, 76; Bignoniaceae, 76; Bixaceae, 76; Bombacaceae, 78; Bonnetiaceae, 82; Boraginaceae, 82; Bretschneideraceae, 83; Bruniaceae, 83; Burseraceae, 83; Buxaceae, 84; Byblidaceae, 84
Cactaceae, 85; Callitrichaceae, 85; Calycanthaceae, 85; Campanulaceae, 85; Canellaceae, 86; Cannabiaceae, 86; Capparidaceae, 86; Caprifoliaceae, 88; Caricaceae, 89; Caryocaraceae, 89; Caryophyllaceae, 90; Casuarinaceae, 91; Celastraceae, 91; Ceratophyllaceae, 95; Cercidiphyllaceae, 95; Chenopodiaceae, 95; Chloranthaceae, 95; Circaeasteraceae, 97; Cistaceae, 97; Clusiaceae, 97; Combretaceae, 103; Compositae, 104; Connaraceae, 105; Convolvulaceae, 110; Coriariaceae, 111; Cornaceae, 111; Corynocarpaceae, 111; Crassulaceae, 111; Crossosomataceae, 111; Cruciferae, 111; Cucurbitaceae, 112; Cunoniaceae, 115; Cynocrambaceae, 115
Daphniphyllaceae, 116; Datiscaceae, 116; Degeneriaceae, 116; Didiereaceae, 117; Dilleniaceae, 117; Dioncophyllaceae, 119; Dipsacaceae, 119; Dipterocarpaceae, 119; Droseraceae, 121

Ebenaceae, 122; Elaeagnaceae, 124; Elaeocarpaceae, 124; Elatinaceae, 128; Ericales, 128; Erythroxylaceae, 128; Escalloniaceae, 129; Eucommiaceae, 129; Euphorbiaceae, 129; Eupomatiaceae, 143; Eupteleaceae, 143

Fagaceae, 143; Flacourtiaceae, 143; Fouquieraceae, 147; Frankeniaceae, 147; Fumariaceae, 147

Garryaceae, 147; Gentianaceae, 148; Geraniaceae, 148; Gesneriaceae, 149; Glaucidiaceae, 149; Gonystylaceae, 149; Goodeniaceae, 150; Grossulariaceae, 151

Halorrhagaceae, 151; Hamamelidaceae, 151; Hernandiaceae, 152; Hippocastanaceae, 152; Hydnoraceae, 153; Hydrangeaceae, 153; Hydrophyllaceae, 153; Hypericaceae, 153

Icacinaceae, 154; Illiciaceae, 154; Ixonanthaceae, 154

Juglandaceae, 155; Julianaceae, 156

Krameriaceae, 156

Labiatae, 156; Lacistemaceae, 156; Lactoridaceae, 156; Lardizabalaceae, 156; Lauraceae, 157; Lecythidaceae, 159; Legnotidaceae, 161; Leguminosae, 161; Leitneriaceae, 173; Lentibulariaceae, 173; Limnanthaceae, 173; Linaceae, 173; Loasaceae, 174; Lobeliaceae, 175; Loganiaceae, 175; Lythraceae, 176

Magnoliaceae, 177; Malpighiaceae, 179; Malvaceae, 180; Marcgraviaceae, 182; Martyniaceae, 182; Melastomataceae, 182; Meliaceae, 185; Melianthaceae, 193; Menispermaceae, 193; Monimiaceae, 194; Moraceae, 197; Moringaceae, 197; Myricaceae, 198; Myristicaceae, 198; Myrsinaceae, 202; Myrtaceae, 202

Nandinaceae, 205; Nepenthaceae, 206; Nyctaginaceae, 206; Nymphaeaceae, 207; Nyssaceae, 207

Ochnaceae, 208; Olacaceae, 209; Oleaceae, 209; Onagraceae, 209; Opiliaceae, 210; Orobanchaceae, 210; Oxalidaceae, 210

Paeoniaceae, 211; Pandaceae, 212; Papaveraceae, 212; Parnassiaceae, 215; Passifloraceae, 215; Pedaliaceae, 216; Phrymaceae, 216; Phytolaccaceae, 217; Piperaceae, 217; Pittosporaceae, 218; Plantaginaceae, 218; Platanaceae, 218; Plumbaginaceae, 218; Podophyllaceae, 218; Podostemaceae, 219; Polemoniaceae, 220; Polygalaceae, 220; Polygonaceae, 222; Portulacaceae, 222; Primulaceae, 222; Proteaceae, 222; Punicaceae, 224

Rafflesiaceae, 224; Ranunculaceae, 224; Resedaceae, 226; Rhamnaceae, 226; Rhizophoraceae, 227; Rosaceae, 228; Rubiaceae, 231; Rutaceae, 232

Salicaceae, 237; Salvadoraceae, 237; Santalales, 237; Sapindaceae, 238; Sapotaceae, 248; Sarraceniaceae, 249; Saururaceae, 249; Sauvagesiaceae, 249; Saxifragaceae, 250; Schisandraceae, 250; Scrophulariaceae, 250; Scyphostegiaceae, 251; Scytopetalaceae, 252; Selaginaceae, 252; Simaroubaceae, 252; Solanaceae, 254; Sonneratiaceae, 255; Sphenocleaceae, 256; Stachyuraceae, 256; Stackhousiaceae, 256; Staphyleaceae, 256; Sterculiaceae, 258; Stylidiaceae, 265; Styracaceae, 265; Symplocaceae, 265

Tamaricaceae, 265; Theaceae, 265; Thymelaeaceae, 270; Tiliaceae, 271; Tovariaceae, 274; Trapaceae, 274; Tremandraceae, 274; Trigoniaceae, 274; Trochodendraceae, 274; Tropaeolaceae, 275; Turneraceae, 275

Ulmaceae, 275; Umbelliferae, 275; Urticaceae, 276

Valerianaceae, 276; Verbenaceae, 276; Violaceae, 276; Vitaceae, 277; Vochysiaceae, 280

Winteraceae, 280

Zygophyllaceae, 282

References   284

Index   305

STANDARD ABBREVIATIONS
FOR FIGURES

a.   aril
c.   cotyledon
e.   endosperm
enc.   endocarp
epc.   epicarp
exc.   exocarp
i.e.   inner epidermis
i.h.   inner hypodermis
i.i.   inner integument
l.s.   longitudinal section
mc.   mesocarp
n.   nucellus
o.e.   outer epidermis
o.h.   outer hypodermis
o.i.   outer integument
p.   pericarp
r.   radicle
rec.   receptacle
t.s.   transverse section

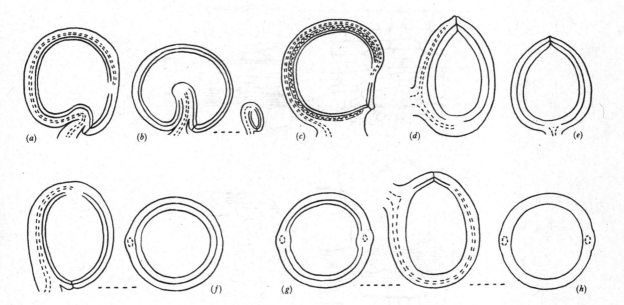

1 Seed-forms derived from the anatropous ovule, except (*d*) (hemianatropous) and (*e*) (orthotropous), showing testa, tegmen, and v.b. (*a*) the obcampylotropous with exaggerated raphe; (*b*) the campylotropous with exaggerated antiraphe; (*c*) the hilar seed with exaggerated hilum extending round most of the seed (as shown by the tracheid bar in *Mucuna*); (*d*) the preraphe seed; (*e*) the orthotropous seed; (*f*) the anatropous seed with t.s.; (*g*) the perichalazal seed in t.s. and (*h*) the pachychalazal seed in t.s., both with similar l.s. (with free integuments at the micropylar end).

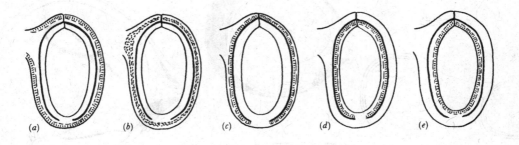

2 Diagrams of the main kinds of seed-structure, showing the testa and tegmen for an anatropous seed. (*a*) exotestal; (*b*) mesotestal; (*c*) endotestal; (*d*) exotegmic; (*e*) endotegmic.

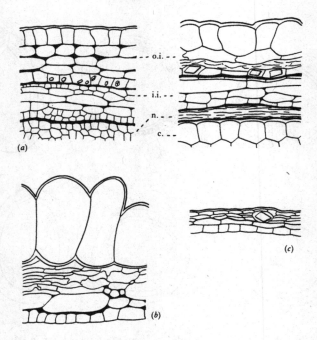

3 *Acer*, seed-coats in t.s., × 300 (after Guérin 1901). (*a*) *A. pseudoplatanus*, immature (left) with endotestal crystals; c. cotyledon. (*b*) *A. pennsylvanicum*. (*c*) *A. negundo*.

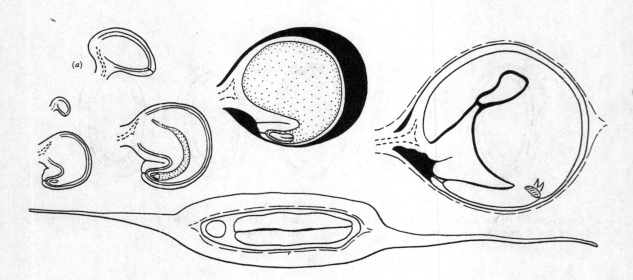

4 *Dipteronia sinensis*, (*a*) ovule in l.s. soon after fertilization, × 25. Developing seeds in median l.s., mature seed (right), × 8. Fruit with seed in t.s., with thin woody endocarp, × 8.

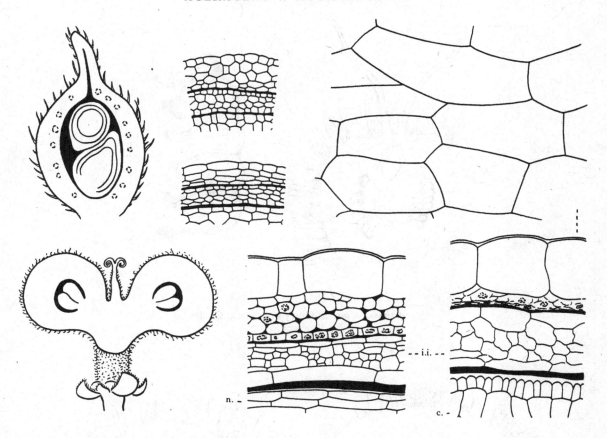

5 *Dipteronia sinensis*. Young fruit shortly after fertilization in l.s., × 8; in t.s. of the follicle, × 25. Wall of young seeds soon after fertilization, of fully grown but immature seed, and of mature seed (lower right), × 200.

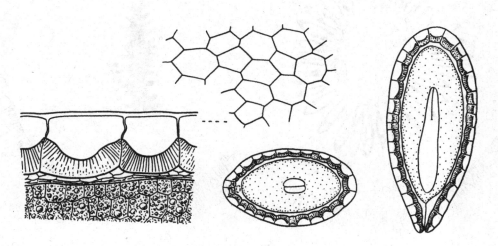

6 *Actinidia chinensis*. Seed in l.s. and t.s., × 25. Seed-coat with endosperm in section, × 120; o.e. facets, × 50.

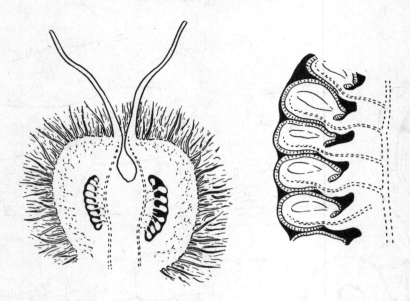

7   *Actinidia chinensis.* Ovary in l.s., × 5. Ovules
in l.s., × 25.

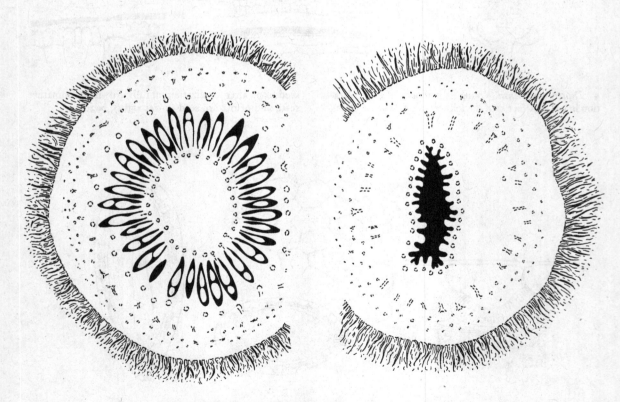

8   *Actinidia chinensis.* Ovary in t.s. near the base and
(right) at the base of the styles, × 10.

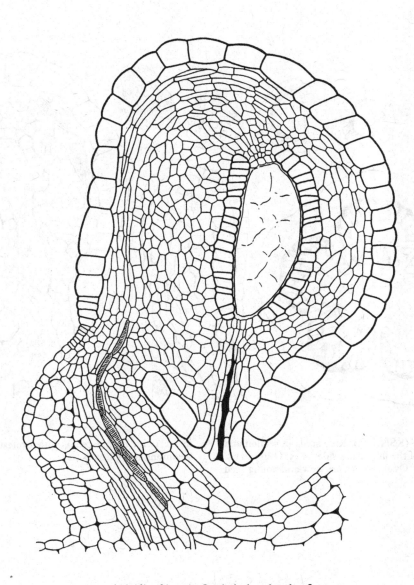

9 *Actinidia chinensis*. Ovule in l.s. shortly after
fertilization, × 225.

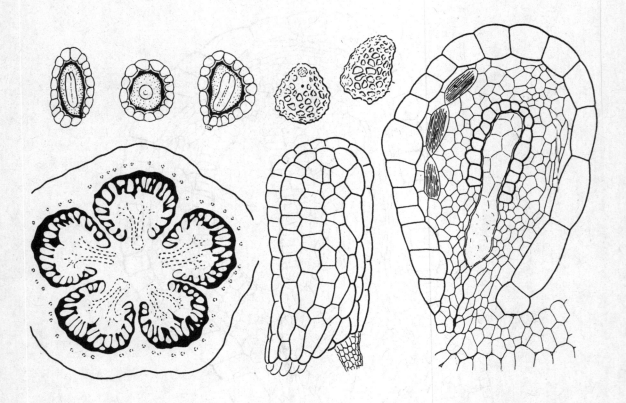

10   *Saurauia sp.* (RSNB 75). Ripe seeds in section and in surface-view of the hilum and side, ×25. Ovary in t.s. at anthesis, ×18. Ovule in surface-view and young seed in l.s. (with raphid-cells), with exotesta and endotesta beginning to lignify, ×225.

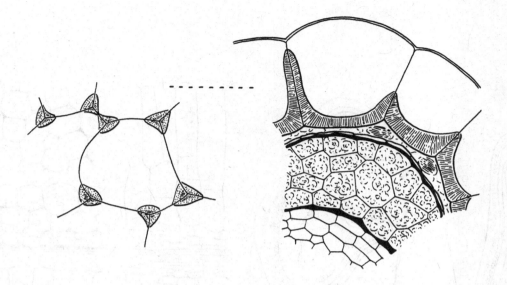

11 *Saurauia sp.* (RSNB 75). Ripe seed in t.s. with exotesta, raphid-cells, trace of endotesta, endosperm, and cotyledon-tissue, with facets of the exotesta, × 225.

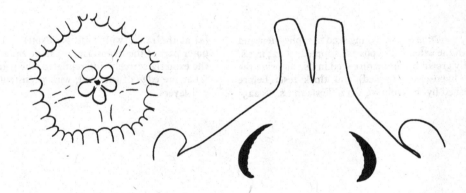

12 *Saurauia sp.* (RSNB 75). Ovary of very young flower-bud (left) with staminal ring and carpels just forming, × 50. Ovary and stamens of an older bud in l.s., × 50.

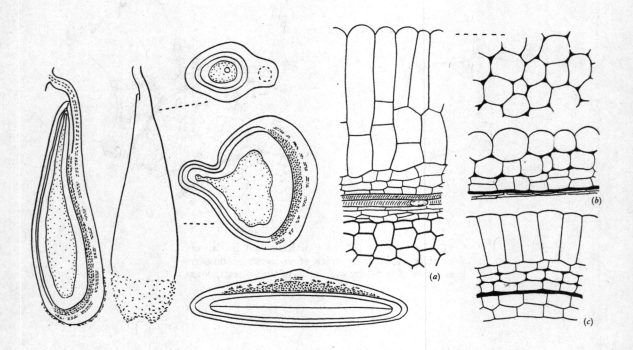

13 *Campnosperma minor*. Young seed in raphe-view and in l.s., with the extended hypostase stippled, × 12; in t.s. × 25. Fully grown but immature seed in t.s., showing the extensive hypostase (stippled) and thick testa before being crushed by the embryo, × 12. Testa in t.s., × 225; (*a*) at the raphe with multiple exotesta, tracheids, and outer part of the hypostase; (*b*) along the antiraphe, with the exotesta scarcely differentiated, the tegmen crushed; (*c*) at the edge of the seed, with persistent tegmen of 2 cell-layers.

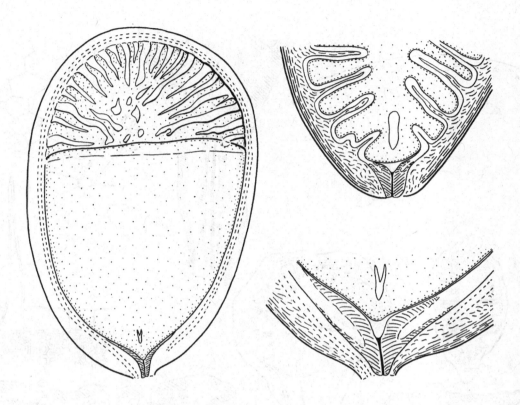

14 *Cyathocalyx carinatus.* Seed in median l.s., × 5.
Micropylar end in median and transmedian l.s., × 10.
Woody tissue of the endostome hatched.

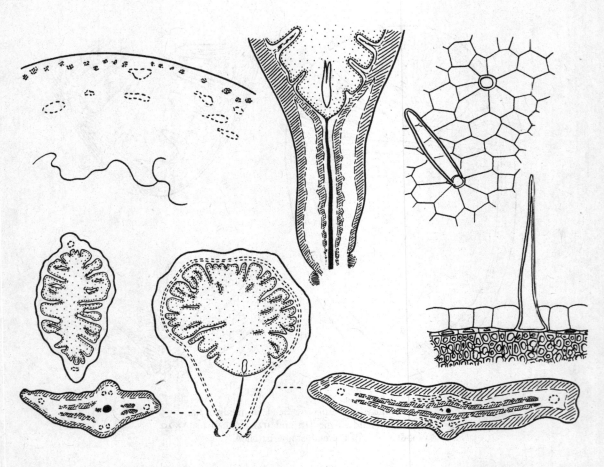

15 *Goniothalamus sp.* (RSS 2302). Seed in l.s. and t.s., ×2; micropylar end in t.s., ×5; in transmedian l.s., ×10; with the fibrous tissue hatched. Surface of the testa with hairs, ×225. Pericarp in t.s., ×5 (upper left).

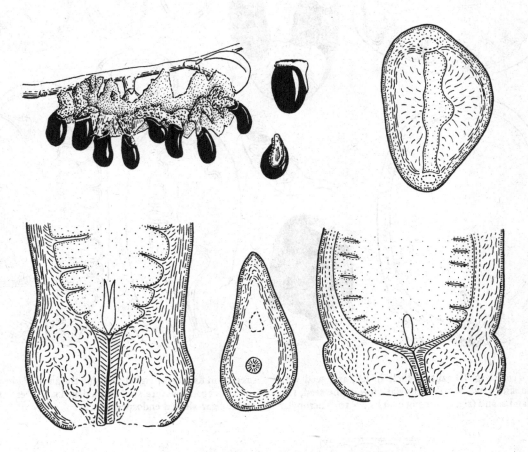

16 *Xylopia peekelii*. Fruits, one dehisced, × ½. Seed in
t.s. and median l.s., × 5; in transmedian l.s. (lower left),
× 10. Woody tissue of the endostome hatched.

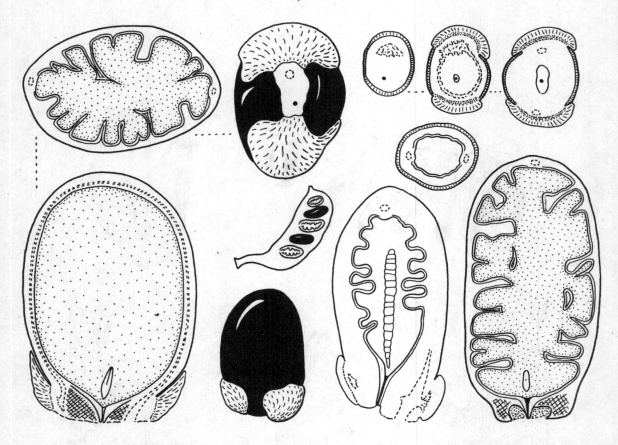

17   *Xylopia sp.* (Brazil). Follicle in l.s., and seed with aril in side-view, × 5. Micropylar view of the seed, t.s., median l.s. and (right) transmedian l.s., × 10. Micropylar region in t.s. (upper right), showing the annular plexus of v.b., × 10. Immature seed in median l.s. (lower centre) with cigar-shaped endosperm, × 25.

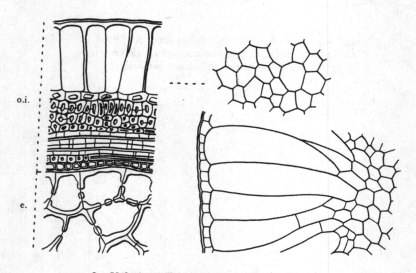

18   *Xylopia sp.* (Brazil). Seed-coat and aril in t.s., × 225.

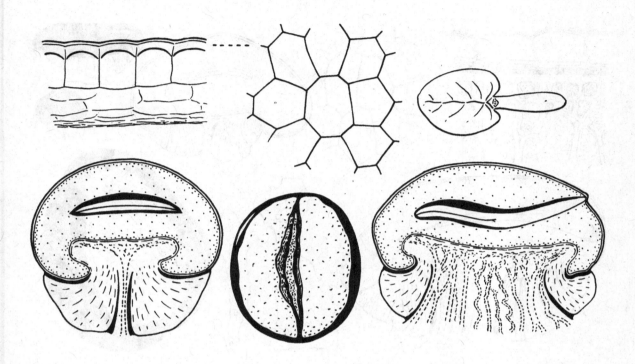

19 *Chilocarpus torulosus*. Seed with aril from the funicular side, ×5. Seed in l.s. and t.s., ×8. Embryo with one cotyledon removed, ×6. Integument in t.s. with epidermal facets, ×225.

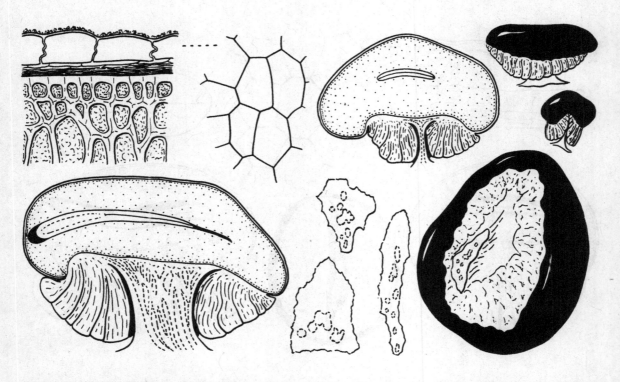

20 *Chilocarpus sp.* (RSNB 8326). Arillate seeds in side-view and end-view, with funicle, × 2. Sections of seed with aril and funicle, and a seed from the underside with the aril removed, × 5. Funicle in t.s., × 10. Seed-coat details with endosperm, × 225.

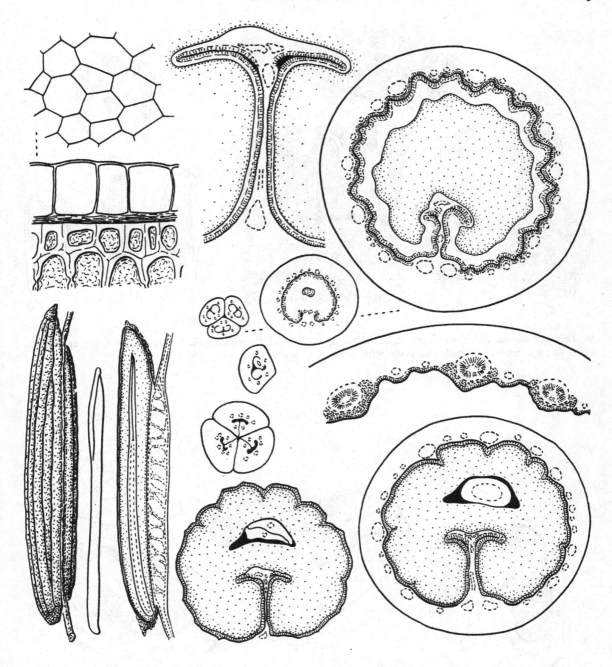

21 *Lepinia solomonensis*. Seeds in side-view and longi-tudinal section with attached placenta, and embryo, × 2. Mature seeds in t.s. (lower row, that on right with pod), × 8. Immature seed in the pods and ovary in t.s., showing the development of the seed-coat and intrusion of the endosperm, × 15. Carpels in t.s. at the base and one carpel with ovules, × 25. Pericarp in t.s., with sclerotic tissue round the larger v.b., × 18. Placental groove of seed, × 25. Seed-coat details with endosperm, × 225.

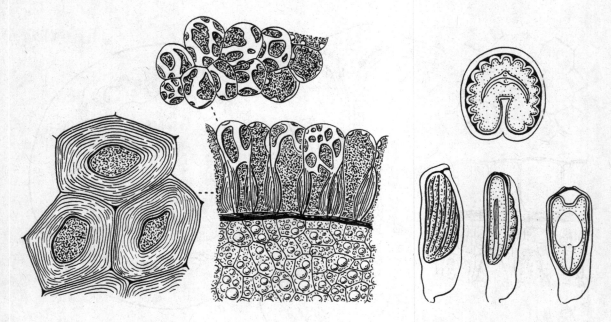

22 *Tabernaemontana dichotoma*. Seeds in l.s. with aril, × 2; in t.s., × 4. Seed-coat details with endosperm, × 75; t.s. of the thick-walled part of the outer epidermis, × 400.

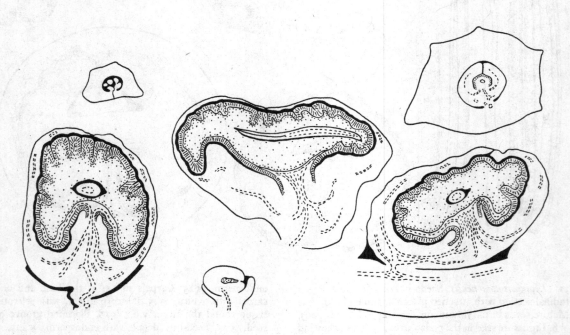

23 *Tabernaemontana sp.* Seeds enclosed in the aril in median l.s. (centre) and t.s., with the exotesta striated, × 8. Carpel in t.s. (upper left), × 12. Young follicle in t.s. (upper right) with vascular aril, × 8. Very young seed in l.s. showing the origin of the aril around the base of the seed, × 12.

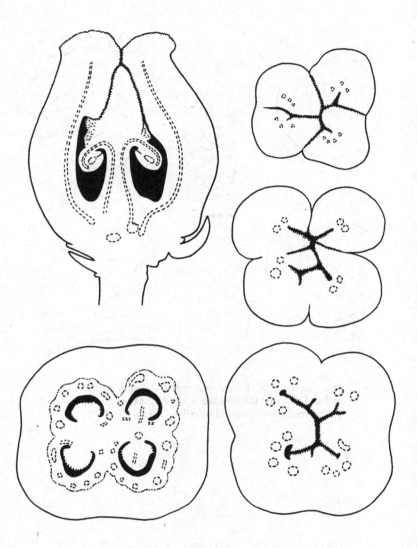

24 *Ilex aquifolia.* Flower in l.s., and in t.s. at various levels from the stigmata to the placentas, v.b. incorporated in the embryonic endocarp, × 18.

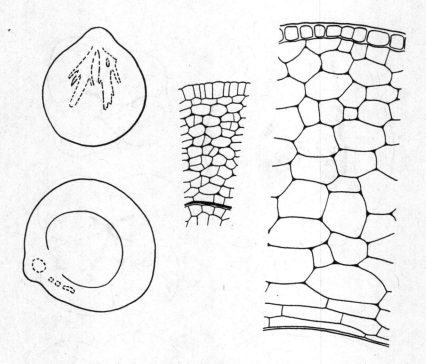

25 *Ilex aquifolia*. Young seed in chalazal view with v.b. and in t.s. across the middle, ×25. Wall of ovule and of young seed with collapsed nucellus, ×225.

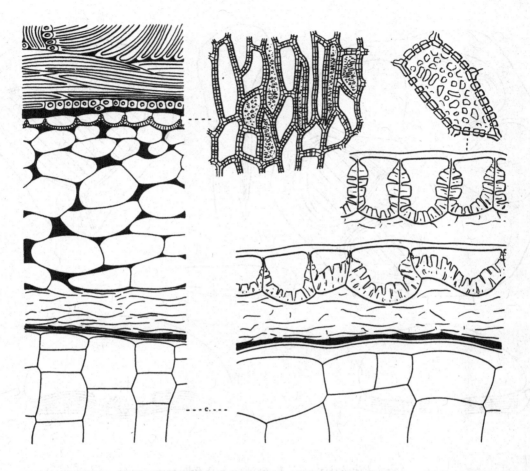

26 *Ilex aquifolia*. Seed-coat in t.s. of a fully grown but immature seed (left), with the endosperm and fibrous endocarp, ×225. Seed-coat of the mature seed (right), ×400.

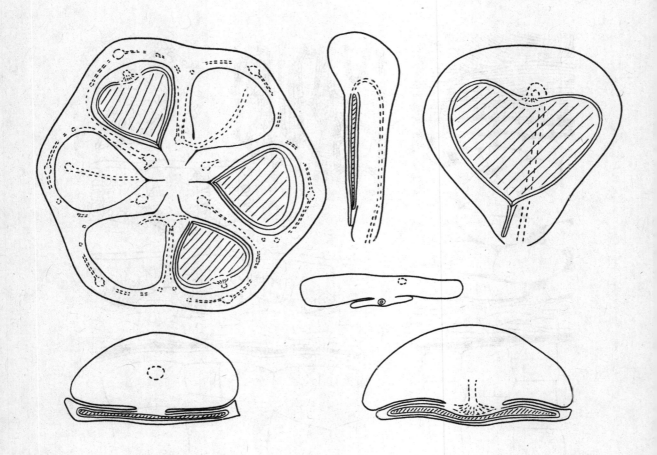

27   *Aristolochia indica.* Fruit in t.s. with fully grown but immature seeds, × 5. Seeds in l.s. and t.s., with the endosperm hatched, × 10.

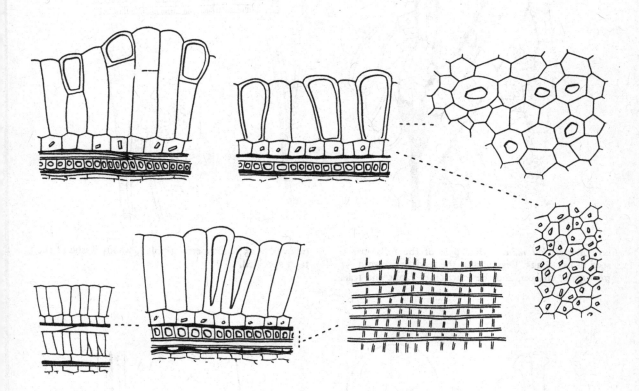

28 *Aristolochia indica*. Seed-coat in l.s. (above) and t.s. (below), with the wall of the young seed (lower left), × 225.

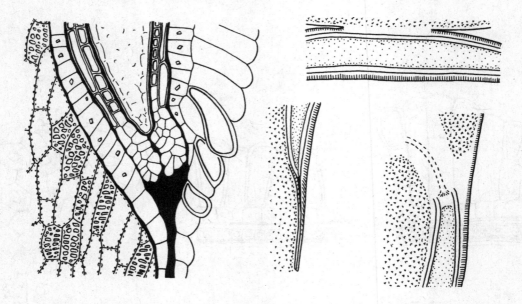

29   *Aristolochia indica.* Micropyle of the fully grown seed in l.s., × 225. Diagrams of the seed-structure at the micropyle, chalaza, and the junction of raphe and testa (above), × 25; endosperm stippled, woody tissue of the testa speckled.

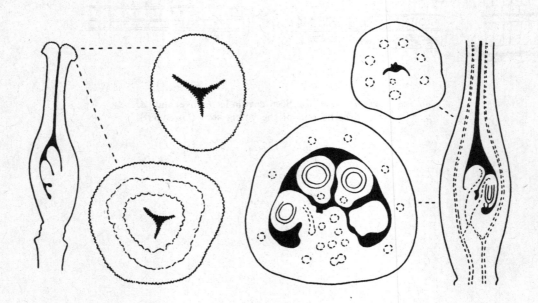

30   *Berberis stenophylla.* Ovary in l.s. and t.s., × 8 (left); and with details, × 25 (right).

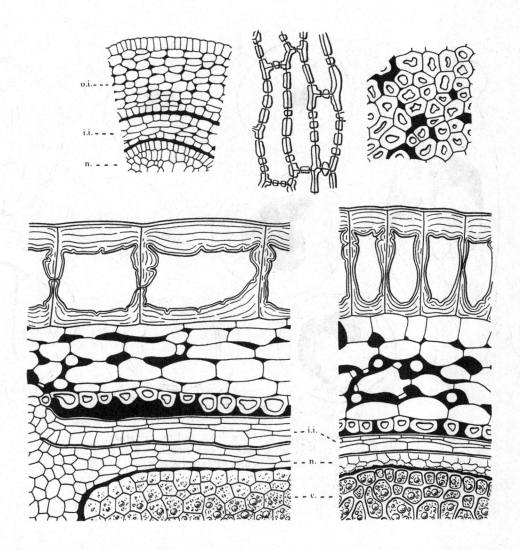

31 *Berberis stenophylla*. Ovule-wall in t.s. and the wall of the fully grown but immature seed in t.s. and l.s. (at the chalazal junction), with exotestal and endotestal facets, × 225.

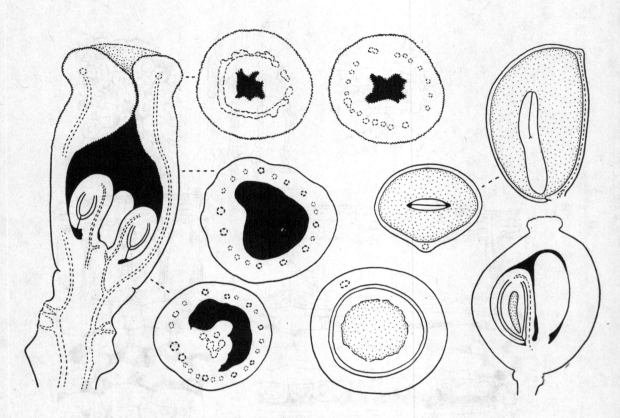

32 *Mahonia japonica.* Ovary in l.s. and t.s. at various levels, ×25. Seed in l.s. and t.s., ×10. Immature seed in t.s., ×25. Immature fruit and seeds in l.s., ×5.

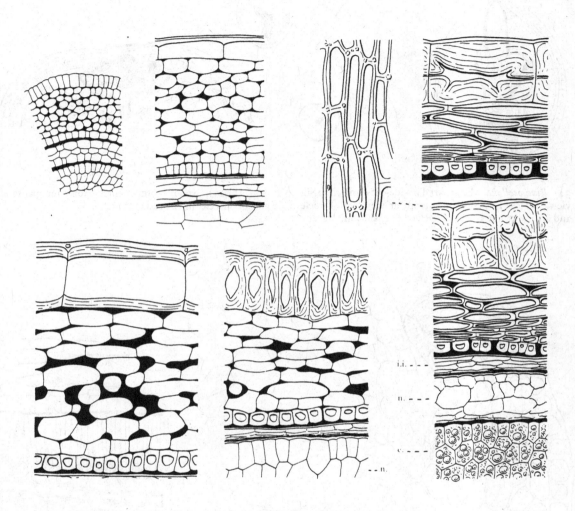

33 *Mahonia japonica*. Wall of ovule and young seed in t.s., of the fully grown but immature seed in t.s. and l.s., and of the mature seed (right) in t.s., ×225.

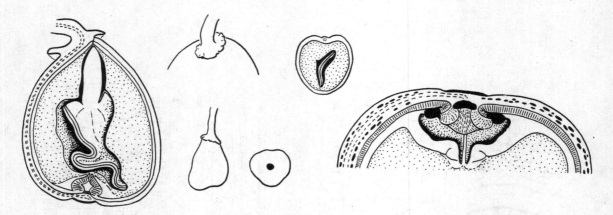

34 *Bixa orellana*. Mature seed in side-view and chalazal, view, × 2; in l.s., × 7; in micropylar view with funicle and aril, × 5; in t.s., × 3. Chalaza in l.s. showing the air- gaps (black), woody components, exotegmic palisade, and bixin-cells in the testa, × 15.

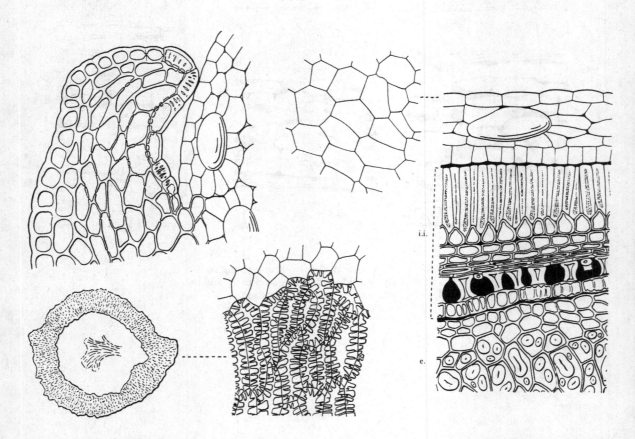

35 *Bixa orellana*. Aril at the junction with the testa, and the mature seed-coat in t.s. (right), × 225. Chalaza in t.s. with the annulus of sclerotic exotestal cells, × 30; facets of the sclerotic exotestal cells, × 225.

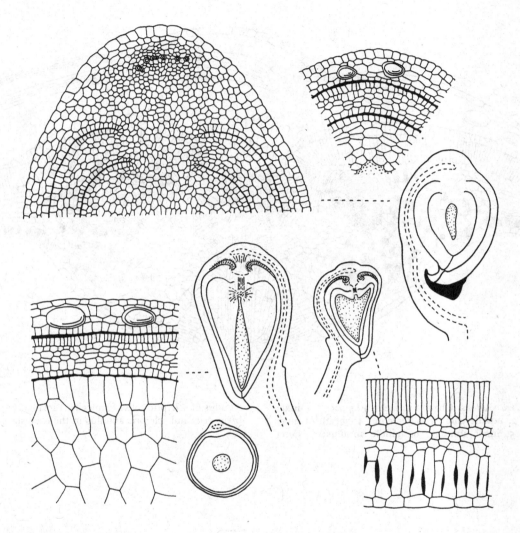

36 *Bixa orellana*. Ovule in l.s. (× 50), with its microscopic structure in transmedian l.s. of the chalaza and in t.s. of the wall (× 225). Young seed with incipient endosperm in l.s. and t.s. (× 15) with the wall in t.s. (× 225). Older seed with more extensive endosperm (× 7), and the incipient tegmen in t.s. (× 225).

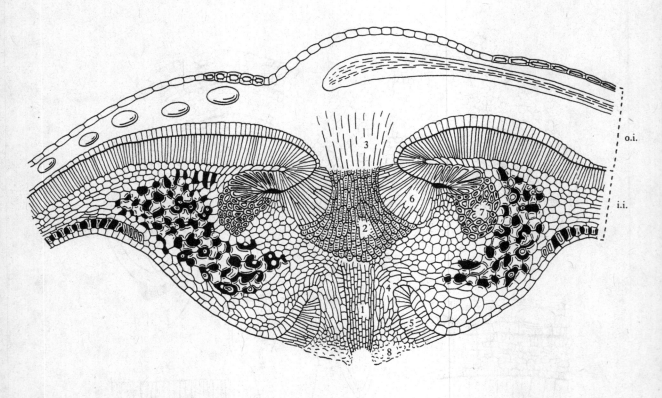

37 *Bixa orellana.* Chalaza in median l.s., ×75. 1, inner core; 2, middle core; 3, outer core; 4, sheath of inner core; 5, inner annulus; 6, middle annulus; 7, outer annulus of sclerotic cells; 8, nucellar remains; testa with bixin-cells and sclerotic annulus in the exotesta.

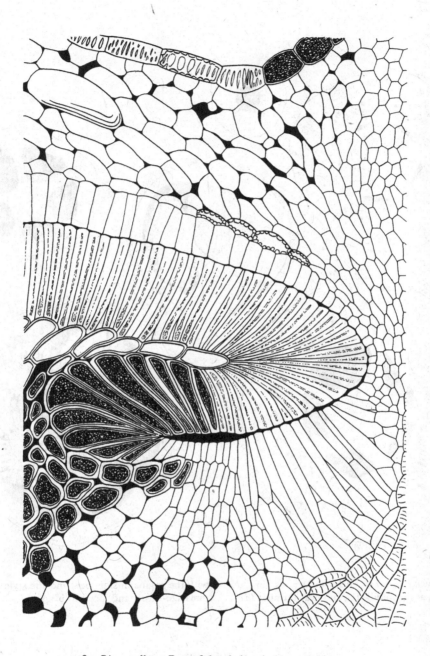

38 *Bixa orellana*. Part of the chalaza in l.s. at the junction of the seed-coats in the fully grown but immature seed; brown gum-cells with dark contents; ×225.

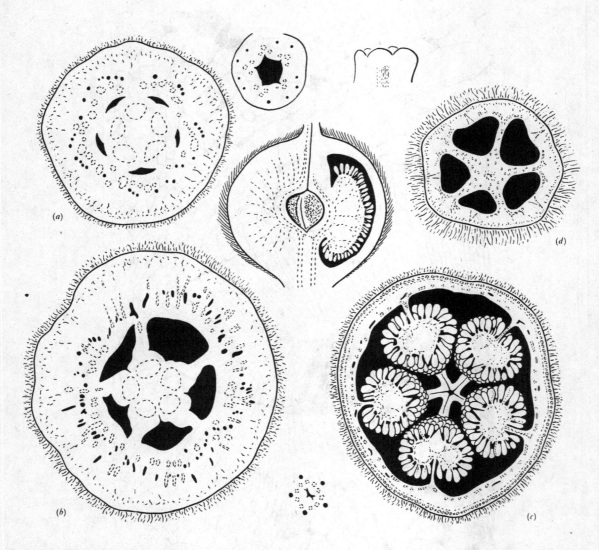

39 *Cochlospermum religiosum.* Ovary in l.s., ×5; in t.s. at the base (*a*, *b*), at the middle (*c*), and near the apex (*d*), with the stigma in side-view and the style in t.s., ×25; mucilage-canals and loculi in black.

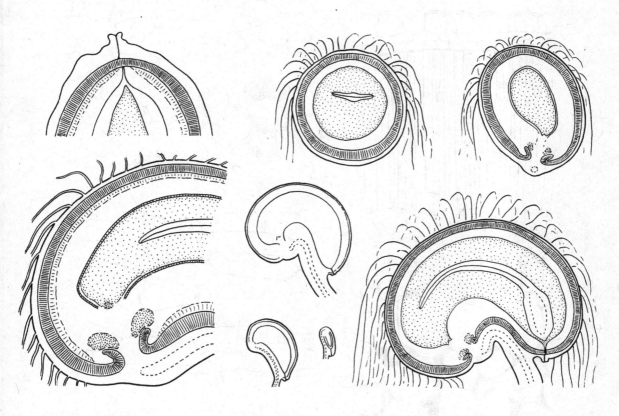

40 *Cochlospermum religiosum*. Ovule and developing seeds in median l.s., × 10. Fully grown but immature seed in l.s. and t.s. (upper right), × 10. Micropyle and chalaza of the fully grown but immature seed, × 25.

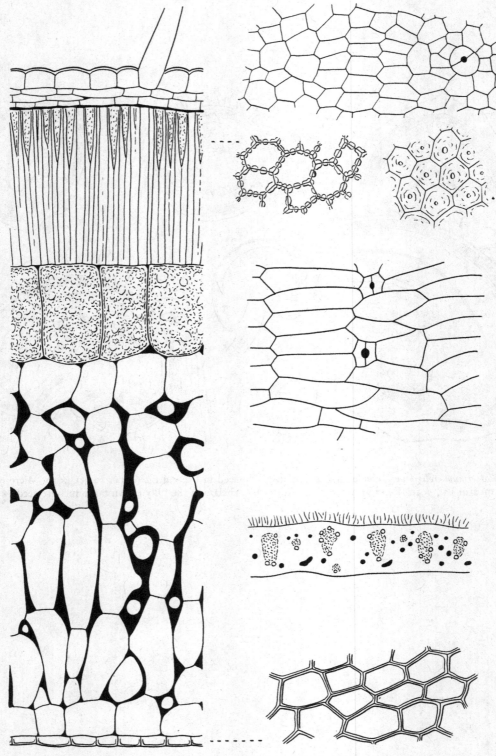

41 *Cochlospermum religiosum*. Seed-coats of the nearly
mature seed in t.s.; testa 3–4 cells thick; tegmen with
palisade (o.e.) and vitreous pigmented cells (o.h.), × 225.
Facets of the exotegmen and endotegmen, and of the
young and mature exotesta with stomata, × 500. Pericarp
in t.s., with the slime-canals in black and the small v.b.
adjacent to the sclerotic patches, × 10.

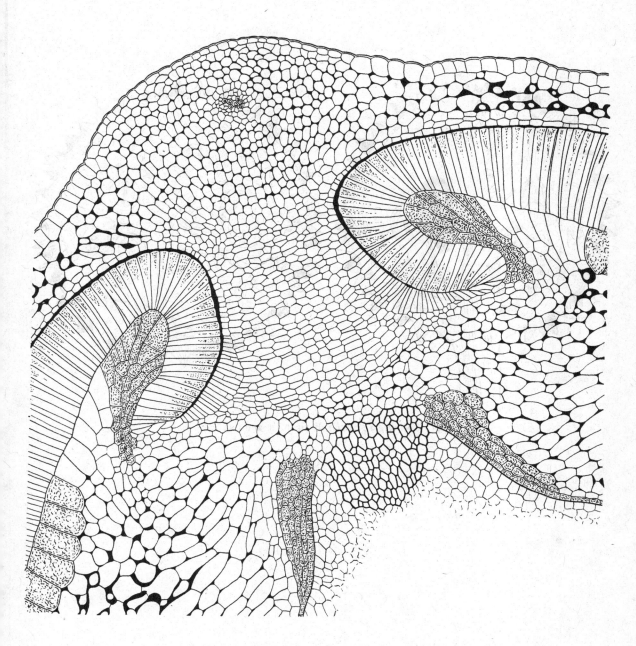

42 *Cochlospermum religiosum*. Chalaza in transmedian
l.s., the pigmented cells with dark contents, × 115.

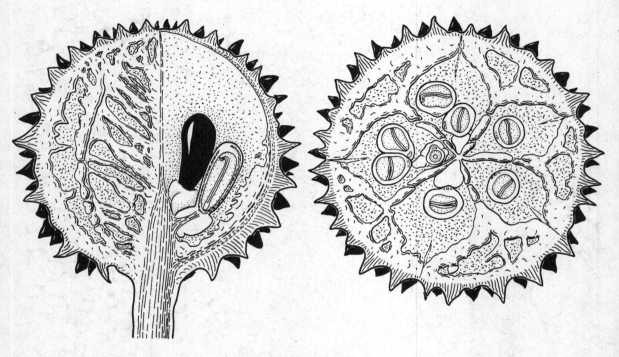

43 *Coelostegia griffithii*. Fruit in l.s. and t.s., × ⅔.

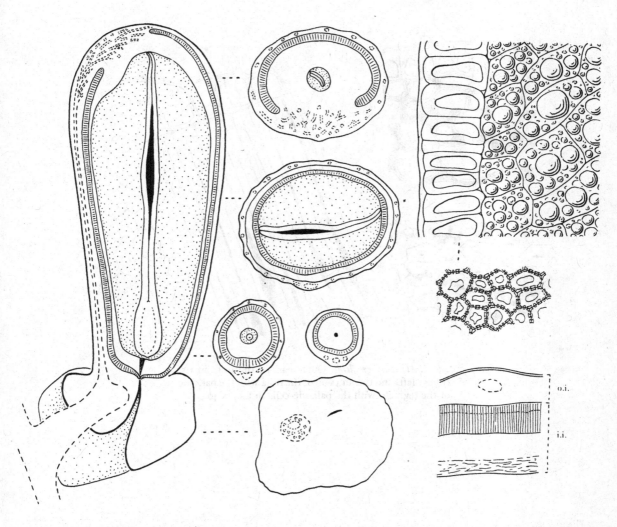

44 *Coelostegia griffithii*. Mature seed in l.s., with t.s. at various levels, × 3. Outer tissue of the aril, × 400. Diagrams of the seed-coats in t.s., the inner part of the tegmen crushed, × 15.

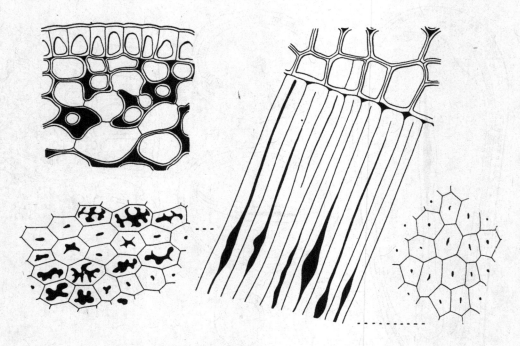

45   *Coelostegia griffithii.* Outer part of the testa in t.s. (upper left) and the junction of the testa with the palisade of the tegmen, with the palisade-cells in t.s., × 400.

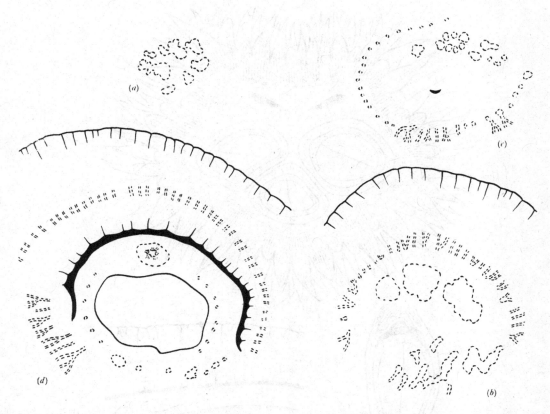

46 *Cullenia zeylanica*. Vascular supply to the seed as seen in t.s., ×8; (a) the initial cluster of v.b. in the funicle; (b) giving off the small v.b. to the aril; (c) at the level of the micropyle (central crescentic slit), with the ring of v.b. for the aril and the remainder consolidating as the compound raphe v.b.; (d) at the beginning of the tegmen, the aril mostly separated from the seed except along the antiraphe.

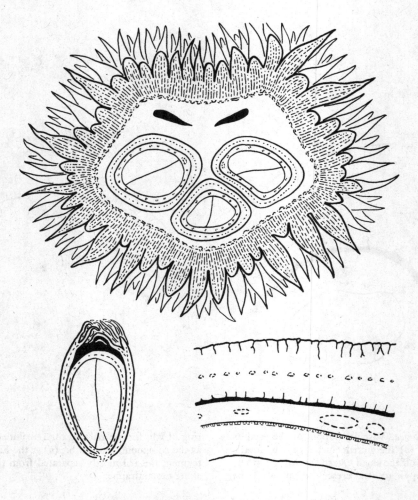

47 *Cullenia zeylanica*. Fruit in t.s., ×1. Seed in l.s.,
×1. Aril and seed-coats in t.s., ×10.

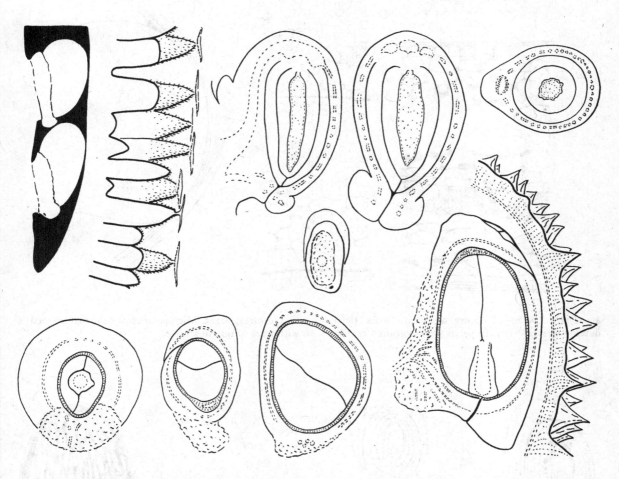

48 *Durio zibethinus*. Upper row; young seeds with the aril beginning, and the adjacent pericarp, ×7; the same seeds in l.s. and t.s., ×13. Centre; the hilar end of a young seed with microphyle and raphe v.b., ×7. Lower row; ripe seeds in t.s. and l.s. (with pericarp), with the seed-coat hatched, and the vascular aril, ×1.

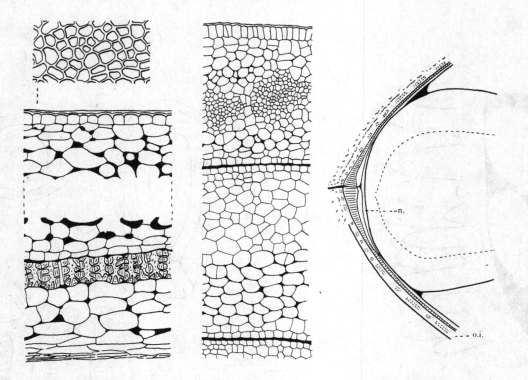

49 *Durio zibethinus*. Mature seed-coats in t.s. (left); developing seed-coats of a young seed soon after fertilisa-tion (centre); × 225. Micropylar end of the mature seed, with trace of nucellus, × 10.

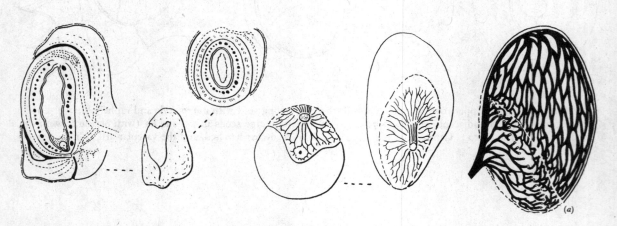

50 *Durio zibethinus*. Young seed with aril, × 1; in l.s. and t.s. with the mucilage-sacs in black, × 7. Hilar views of the seed (aril removed) showing the vascular supply, × 1. Vascular supply of the seed in side-view; *a*, marking the line of detachment of the aril from the testa, × 1.

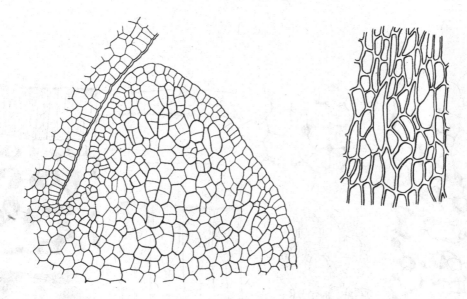

51 *Durio zibethinus*. The very young aril in l.s., and the facets of the mature epidermis of the aril, × 225.

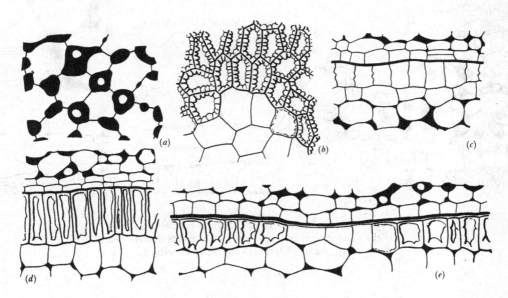

52 *Durio sp.* (S 29470). Mesophyll of the testa (*a*); the tegmic palisade in surface-view (*b*), and a section of the thin-walled part (*e*); × 225. *Durio acutifolius*, section of the tegmic palisade (*d*), × 225. *Cullenia zeylanica*, the thin-walled tegmic palisade in section (*c*), × 225.

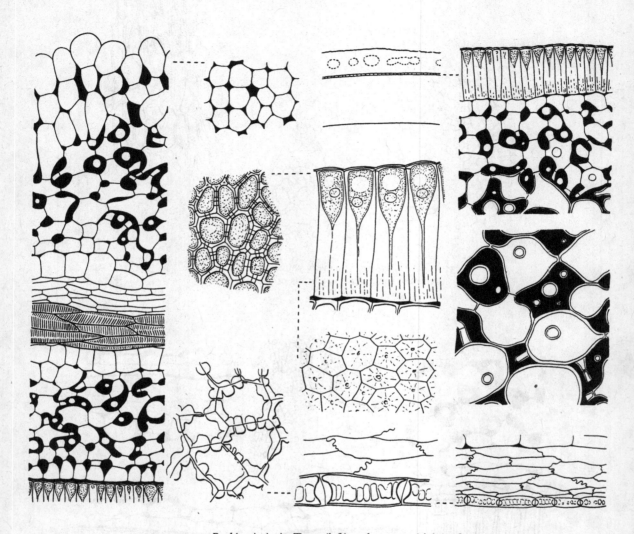

53 *Pachira insignis*. Testa (left) and tegmen (right) of the mature seed in t.s., ×225; with details (except the exotesta), ×400. Diagram of the seed-coats in t.s., ×10.

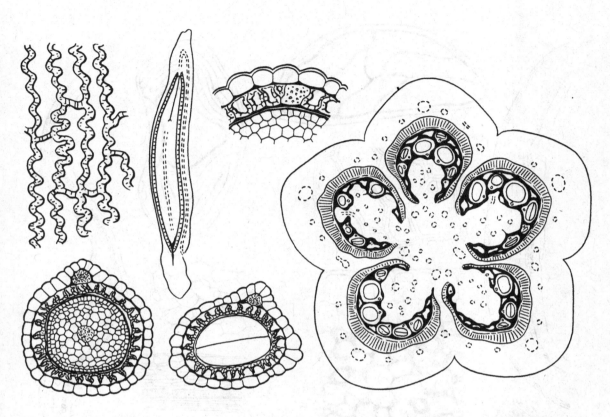

54 *Ploiarium alternifolium.* Capsule in t.s. with the valves of the endocarp (each with a thick outer layer of radiating sclerotic cells and a thin inner layer of transverse sclerotic cells) and v.b., ×18. Seed in l.s., with the woody exotegmen hatched, ×25. Seed in t.s. of the radicle and of the cotyledons, ×115. Seed-coat in t.s., with the exotesta and exotegmen, ×225. Exotegmic cells in surface-view, ×225.

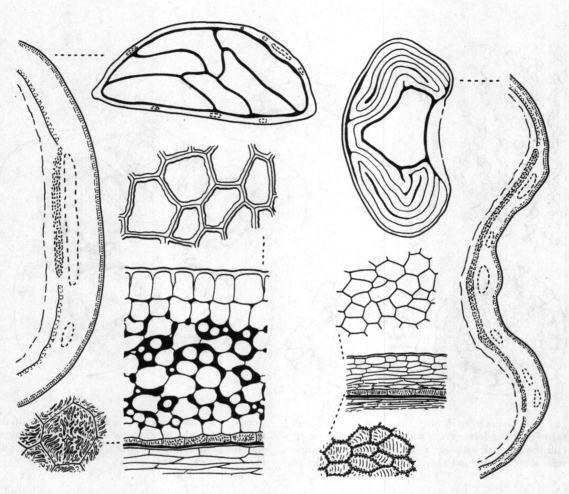

55 *Canarium indicum* (left); seed in t.s., ×5; peri-   ×500. *Commiphora caudata* (right); seed in t.s. at the
chalaza in t.s. with the hypostase stippled, the two epi-   level of the hypocotyl, ×10; perichalaza (? pachycha-
dermal layers of the testa, and the nucellar remains,   laza) in t.s., as in *Canarium*, ×25; testa in t.s., ×225;
×25; testa in t.s., ×225; endotesta in surface-view,   endotesta, ×500.

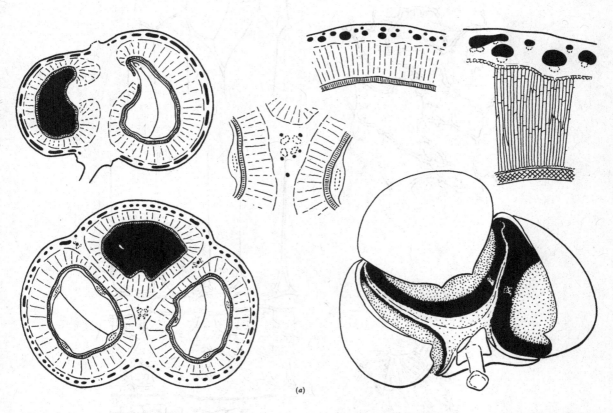

(a)

56 *Protium unifoliolatum*. (a) Fruit in l.s. and t.s., and dehiscing fruit, ×5. Pericarp in t.s., ×10; cell-details, ×25. Axis of fruit in t.s., ×10. Resin-canals and loculi in black. (b) Inner part of pericarp and seed-coats in t.s., ×225; enc. endocarp, mc. mesocarp. Facets of the exotesta (upper left) and the endotesta, ×400.

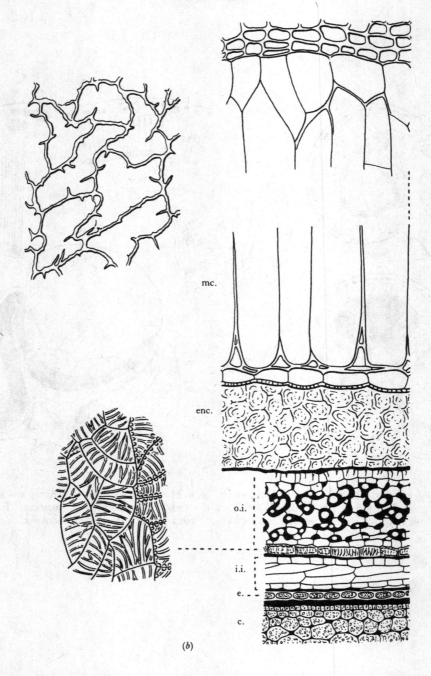

mc.

enc.

o.i.

i.i.

e.

c.

(b)

56 *Protium unifoliolatum*. (*a*) Fruit in l.s. and t.s., and dehiscing fruit, ×5. Pericarp in t.s., ×10; cell-details, ×25. Axis of fruit in t.s., ×10. Resin-canals and loculi in black. (*b*) Inner part of pericarp and seed-coats in t.s., ×225; enc. endocarp, mc. mesocarp. Facets of the exotesta (upper left) and the endotesta, ×400.

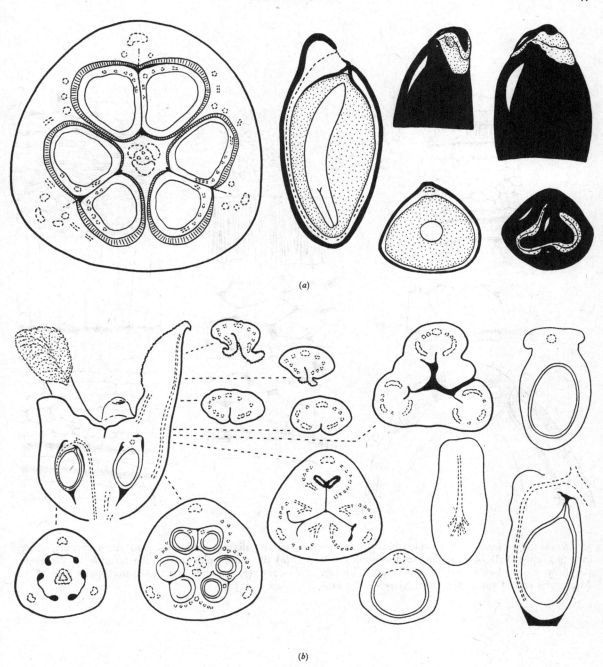

(a)

(b)

57 *Buxus sempervirens.* (a) Fruit with immature seeds in t.s. (endocarp hatched), × 15. Ripe seeds in l.s., t.s., side-view and apical view of the micropylar end (caruncle stippled), × 10. (b) Ovary in l.s. and t.s. at various levels, showing the carpellary construction of the style and stigmata, and the basipetal construction of the ovary, × 15. Ovule (right) in l.s., transmedian l.s., t.s., and raphe-view, × 25.

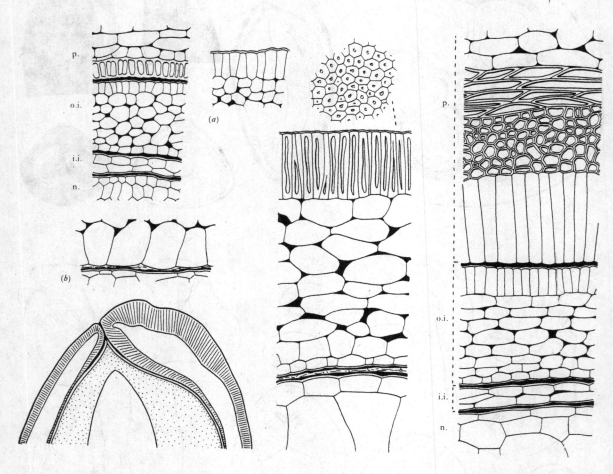

58 *Buxus sempervirens*. Wall of ovule and adjacent pericarp (p.) (upper left), of the mature seed (centre), and of the fully grown but immature seed (right), × 225. Micropylar end of the mature seed, showing the exo-testal palisade continuous into the endotesta, × 25; (*a*) the outer epidermis of the caruncle of the ovule, × 225; (*b*) the endotesta in the micropylar part of the seed, × 225.

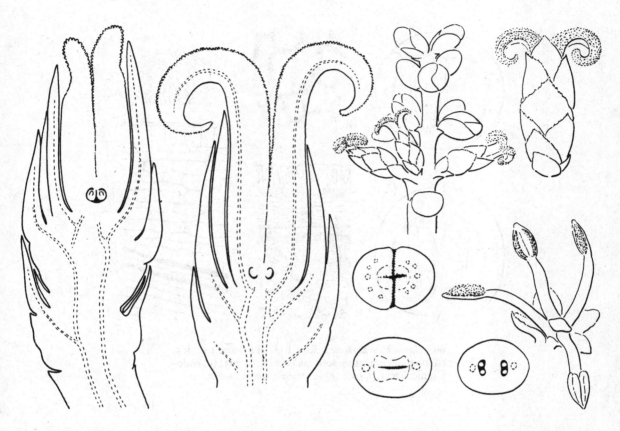

59 *Sarcococca zeylanica*. Female flowers in median and transmedian l.s., with t.s. at three levels, × 25. Axillary inflorescence with basal female flowers and distal male flower-buds, × 5. Female flower, × 10. Male flower, × 5.

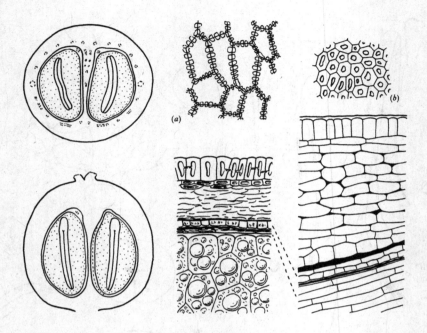

60   *Sarcococca zeylanica.* Ripe fruit in l.s. and t.s., × 5. Mature and immature seed-coats in l.s., × 225; (*a*) endo-tegmic facets, × 500; (*b*) exotegmic facets, × 225.

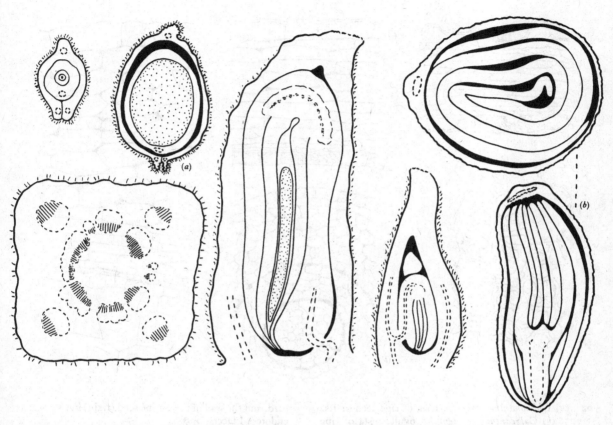

61 *Calycanthus occidentalis*, ovary and young fruit in l.s., showing the broad chalaza, ×25; peduncle in t.s., ×25; (a) ovary and young fruit in t.s., ×10; (b) *Chimonanthus praecox*, ripe seed in t.s. and transmedian l.s., ×10.

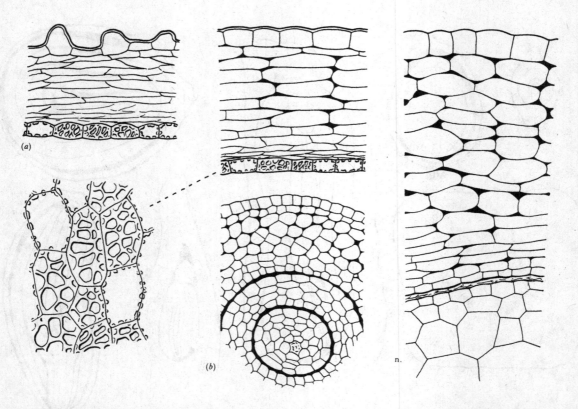

62 (a) *Chimonanthus praecox*, testa of ripe seed in t.s.,   seed, and the wall of the unripe seed (right) in t.s., × 225;
× 225. (b) *Calycanthus occidentalis*, ovule, testa of ripe   endotestal facets, × 500.

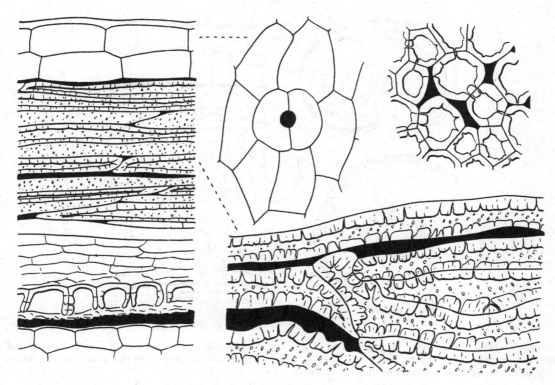

63 *Capparis grandis*. Seed-coats in l.s., exotestal facets with stoma, exotegmic fibres in surface-view, and endo-tegmic facets, (upper right), × 500.

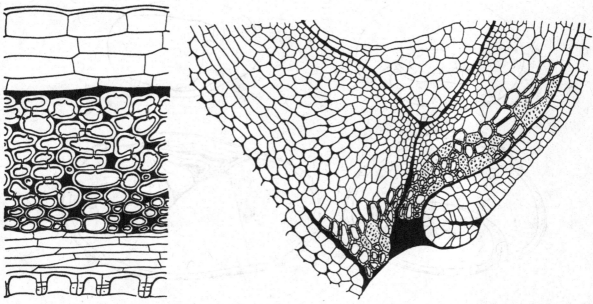

64 *Capparis grandis*. Seed-coats in t.s., with the exo-tegmic fibres cut transversely, × 500. *Capparis divaricata*; micropyle of immature seed in l.s., showing the basipetal lignification of the tegmen, × 115.

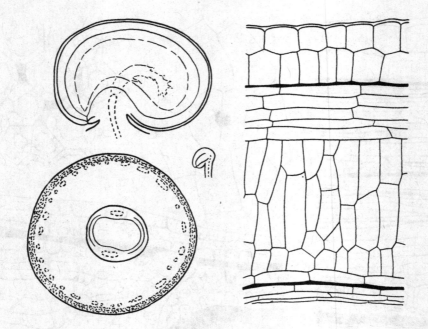

65 *Capparis sepiaria*. Ovule, ×25. Young seed in l.s., showing the integuments, nucellus, watery endosperm, and v.b. of the raphe with a branch to the side of the seed, ×25. Young fruit in t.s. with sclerotic exocarp and young seed, ×10. Seed-coats of a young seed (right) in t.s., ×225.

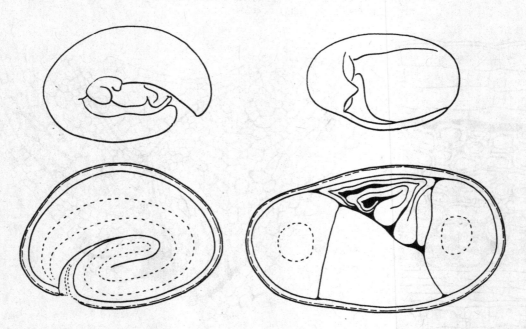

66 *Capparis zeylanica*. Seed in l.s. with long radicle-hypocotyl, and embryo in side-view and under- (or radicle-) view, ×5. Seed in transmedian t.s. (lower right), ×8.

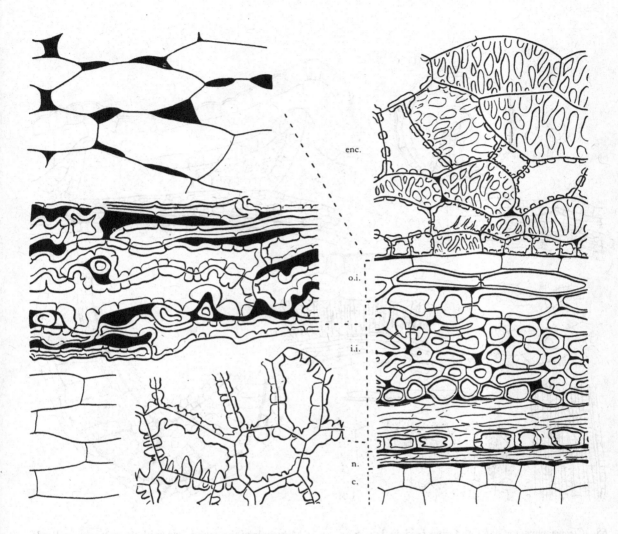

67 *Capparis zeylanica*. Seed-coats with adjacent endo-carp (enc.) and cotyledon (c.) in t.s., with exotestal, exotegmic, and endotegmic facets, and those of the tegmic i.h., × 500.

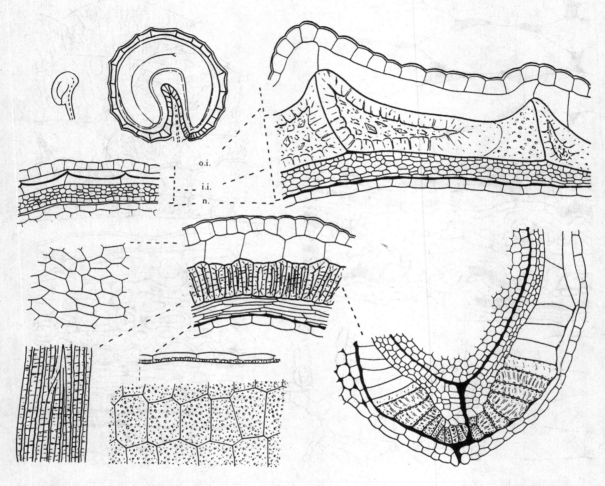

o.i.

i.i.

n.

68　*Cleome viscosa.* Ovule and ripe seed in l.s., × 25.
Immature and mature seed-coats in l.s., × 225. Mature
seed-coat (centre) in t.s., × 225; endotegmic facets × 500.

Micropyle of the young seed, with the exotegmic palisade
lignifying basipetally from the endostome, × 225.

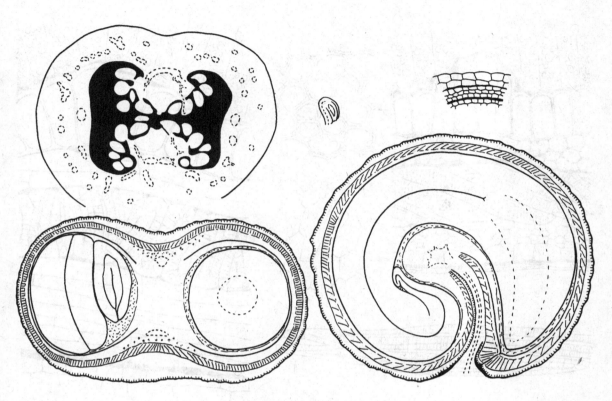

69 *Crataeva religiosa*. Ovary in t.s., × 18. Ovule, × 25.
Ovule-wall in t.s., × 225. Seed in l.s. and t.s., × 18.

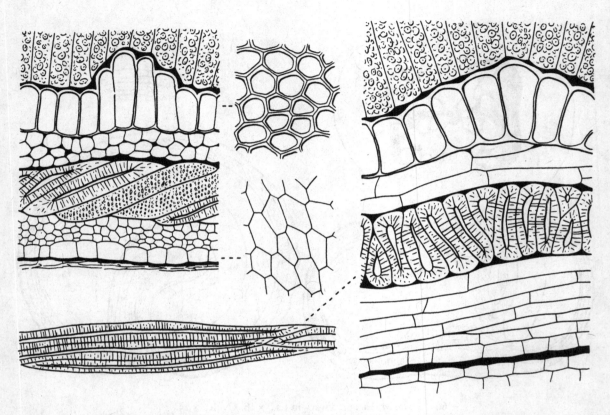

70 *Crataeva religiosa.* Seed-coats in t.s. (left, with collapsed nucellus) and in l.s. (right) with adjacent endocarp (cells filled with starch), × 225; testa 3–4 cells thick;
tegmen with fibrous o.e.

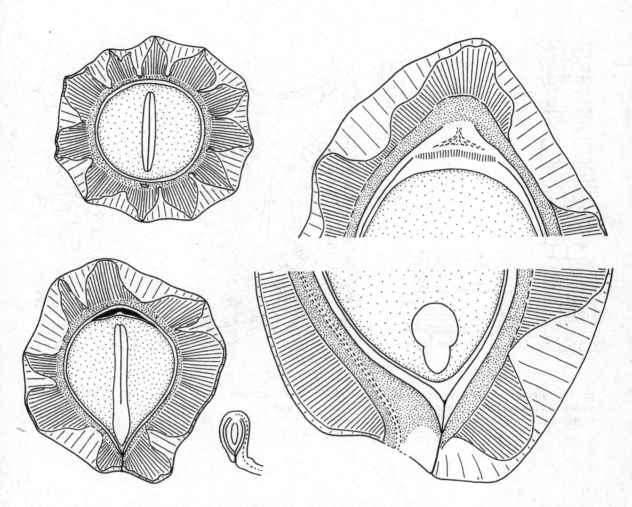

71 *Carica papaya*. Seeds in l.s. and t.s., ovule in l.s., × 10. Chalaza in transmedian l.s. and micropyle in median l.s., from a fully grown but immature seed, × 25.

Pulpy exotesta shown with wide striation; outer mesotesta with close striation; inner mesotesta closely stippled.

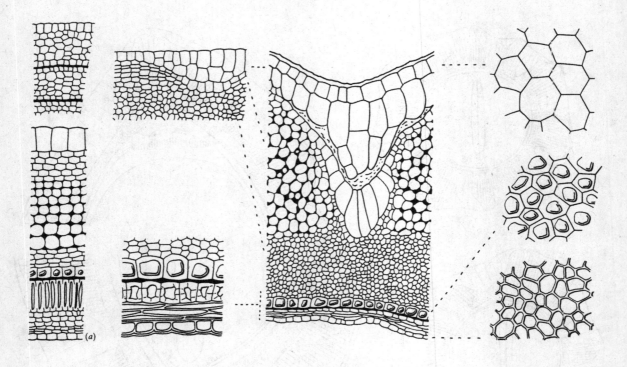

72 *Carica papaya.* Wall of the ovule and of developing seeds in t.s., ×225; mature seed-coats in t.s., with cell-facets, ×115; tegmen in t.s. (after Stephens), ×200.

(*a*) *Jacaratia conica*, seed-coats in t.s. (after Kratzer; mag.).

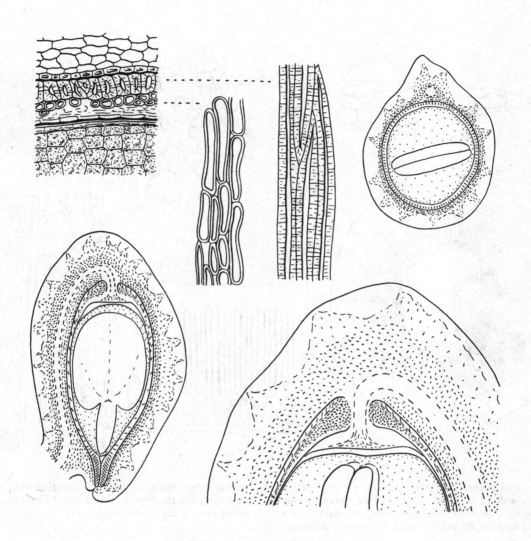

73 *Cylicomorpha parviflora*. Seed in l.s. and t.s., × 10.
Chalaza in l.s., × 25. Tegmen and inner part of the testa,
with endosperm, × 225.

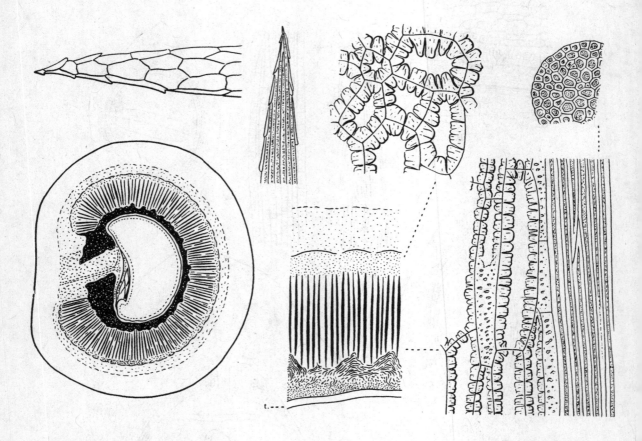

74 *Caryocar butyrosum*. Fruit in l.s., × 1. Part of the pyrene-wall in t.s., with testa (t.), the fibrous-woody endocarp with short irregular ridges, the dark brown woody needles (in black) embedded in the sclerenchymatous inner mesocarp, the loosely sclerenchymatous outer mesocarp, and the firm inner tissue of the exocarp, × 5; cell-details, × 225; needle-tips, × 225.

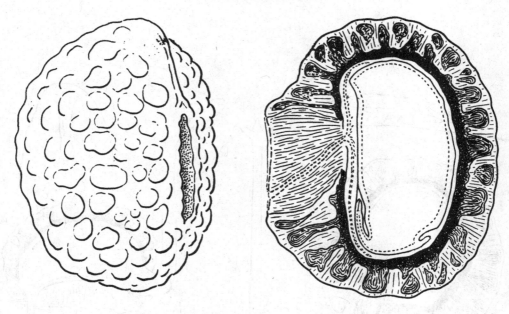

75 *Caryocar nuciferum*. Nut or pyrene in hilar-view and in l.s., × 1; woody tissue striated and stippled.

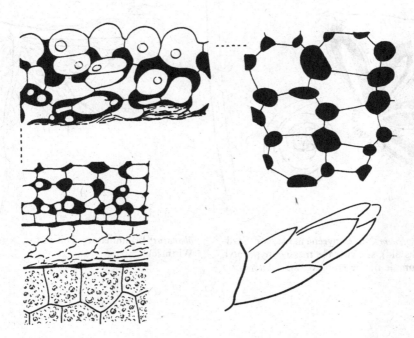

76 *Caryocar butyrosum*. Testa of the mature seed in t.s., wholly aerenchymatous, with the tegmen collapsing, and with the firm tissue of the hypocotyl, × 225. Plumule of the embryo with two basal cotyledons (as scale-leaves), × 5.

(a)

77   *Caryocar nuciferum*. Nut or pyrene in t.s., × 1. Seed-ling (*c.* 11 months old), showing the massive hypocotyl, ×½; inflexed plumule of the embryo, × 1. *Anthodiscus* *obovatus*; (*a*) immature fruit in t.s., × 10; embryo (after Wittmack).

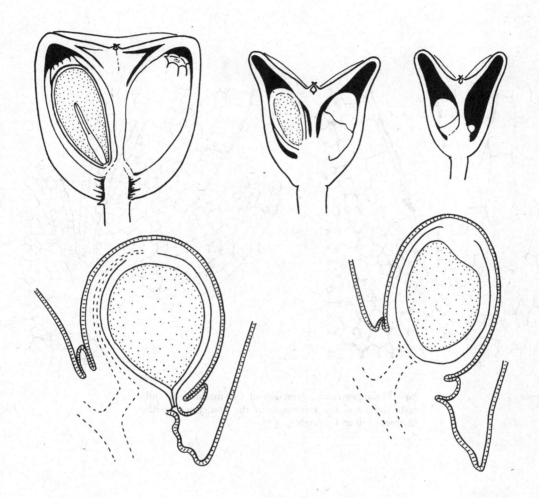

78 *Bhesa paniculata*. Developing fruits in l.s., with ripe fruit (left), × 2. Young seeds in median and transmedian l.s., with incipient aril, × 15; the large-celled epidermis of the testa, aril, and endocarp hatched.

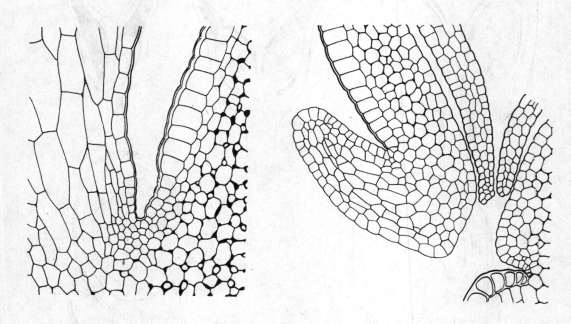

79   *Bhesa paniculata*. Junction of the mature aril and testa (left) and the micropyle of the young seed with incipient aril in l.s. (right), × 75.

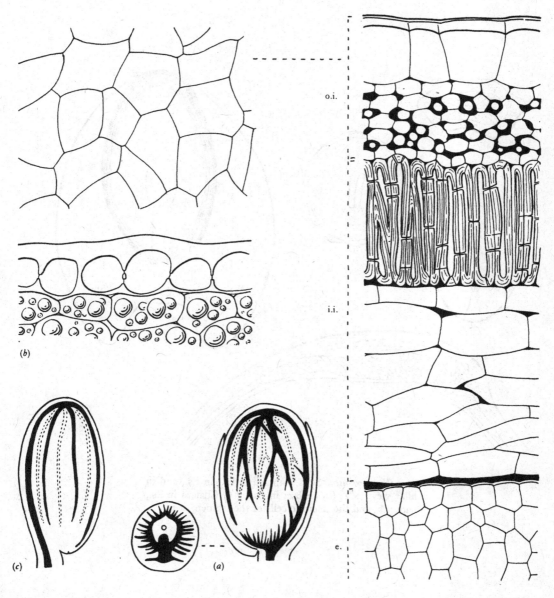

80   *Bhesa paniculata*; mature seed-coat in t.s. (right), × 225; (*a*) vascular supply of the seed, × 3. *Bhesa robusta*; (*b*) epidermis of the aril in t.s., × 225; (*c*) vascular supply of the seed, × 3.

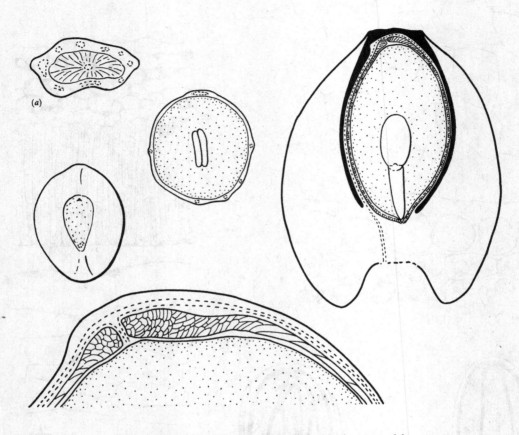

81 *Bhesa robusta*. Seed with aril in l.s., in t.s., and in
hilar view, ×3; (*a*) chalaza in t.s., ×6. Chalaza in l.s.,
with v.b. and the sclerotic cells of the tegmen, ×15.

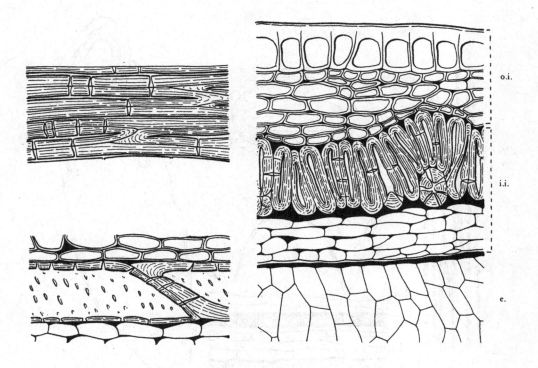

82 *Bhesa robusta.* Seed-coats in t.s., with the fibrous
exotegmen in l.s. (lower left) and surface-view (upper
left), × 225.

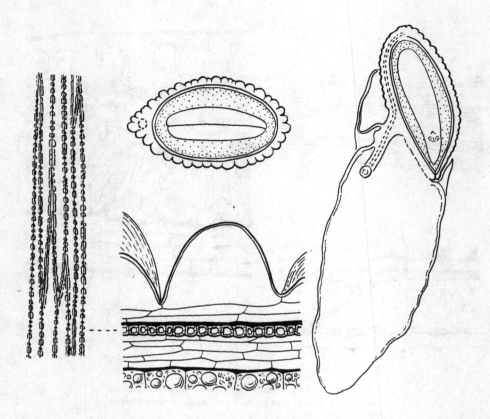

83   *Catha edulis*. Seed in l.s. with funicle and wing-like aril, ×12; in t.s., ×25. Seed-coats in t.s., showing the large papilliform cells of the exotesta and the exotegmic fibres, ×225.

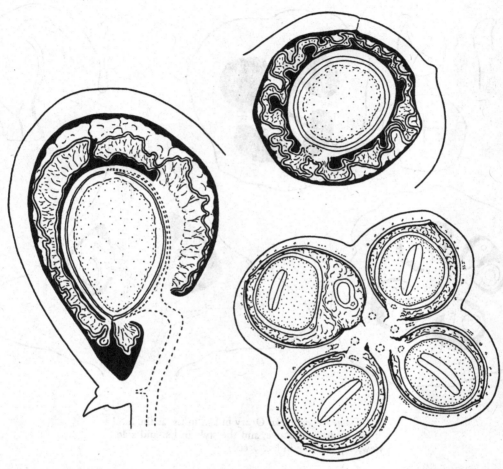

84 *Euonymus europeus.* Mature fruit in t.s., with the septal valves supporting the loculi, × 5. Seeds with pericarp in l.s. and t.s., showing the construction and attachment of the aril (developed from micropyle and raphe), × 10.

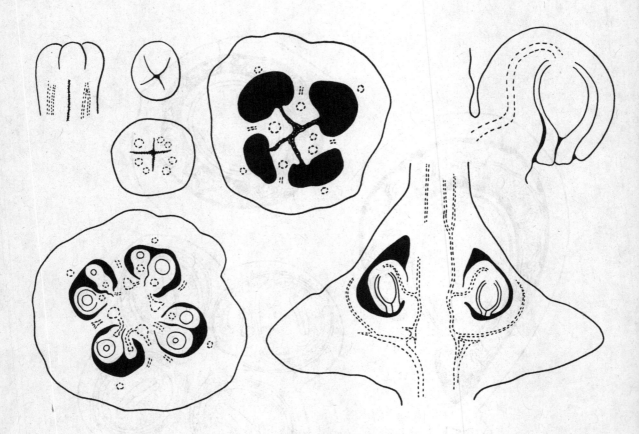

85   *Euonymus europeus.* Ovary in l.s., in t.s. at the level of the ovules and above, and the style in t.s. and side-view, ×25. Ovule in l.s., ×60.

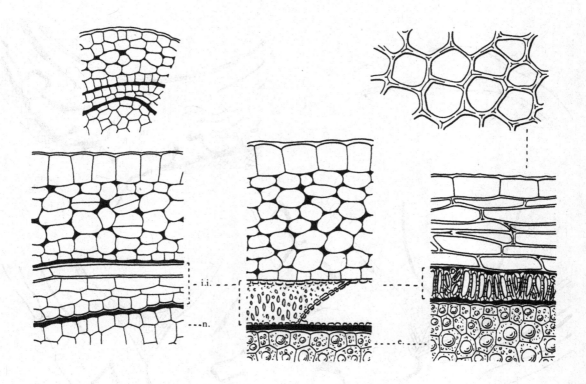

86 *Euonymus europeus.* Wall of ovule, of immature seed (lower left) and of mature seed in l.s. and t.s., × 225; tegmen becoming reduced to the fibrous exotegmen.

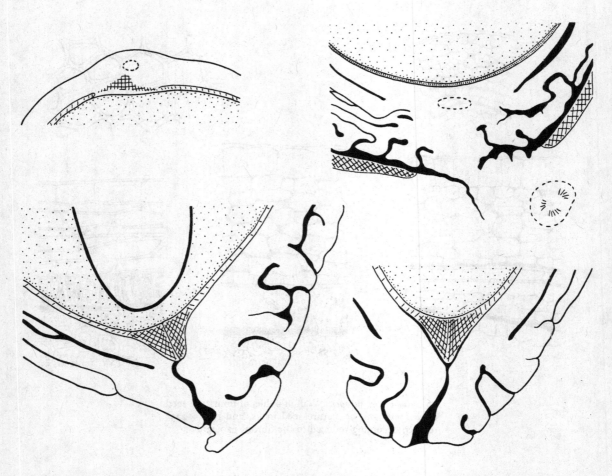

87 *Euonymus europeus*. Seed-structure; ×25. Above, the chalaza in transmedian l.s., with the tracheidal tissue hatched, and the raphe in t.s. showing the aril-attachment, placental v.b. and endocarp (hatched). Micropyle and arillostome in median and transmedian l.s. Tegmen shown with striation.

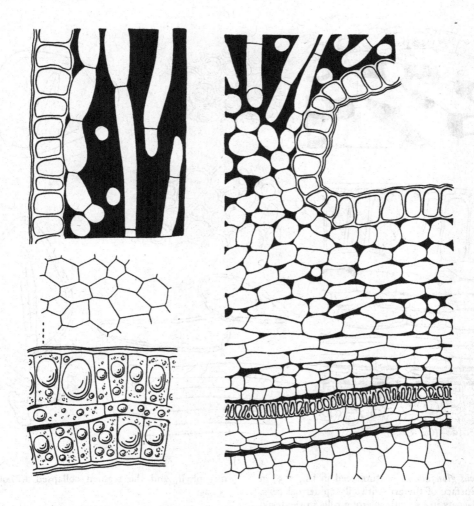

88 *Euonymus europeus.* Aril-structure, × 225. Junction of aril and testa (right) in the fully grown but immature seed with intact tegmen; outer part of the immature aril (upper left); mature aril-lobes in t.s. (lower left).

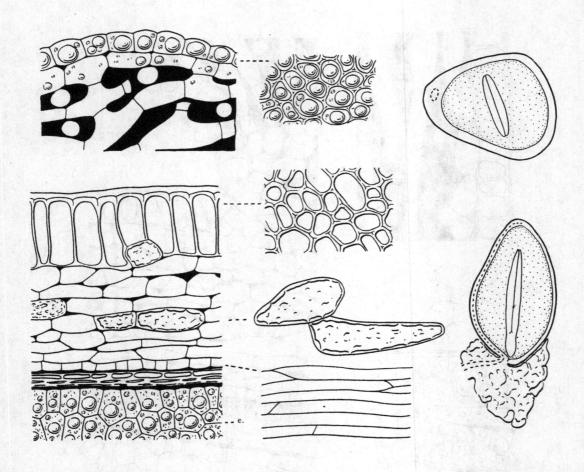

89 *Euonymus glandulosus*. Mature seed in l.s., × 5; in t.s., × 10. Surface of the aril with oily epidermal cells, × 225. Seed-coats in l.s. with sclerotic cells in the testal mesophyll, and the tegmen collapsed without fibres, × 225.

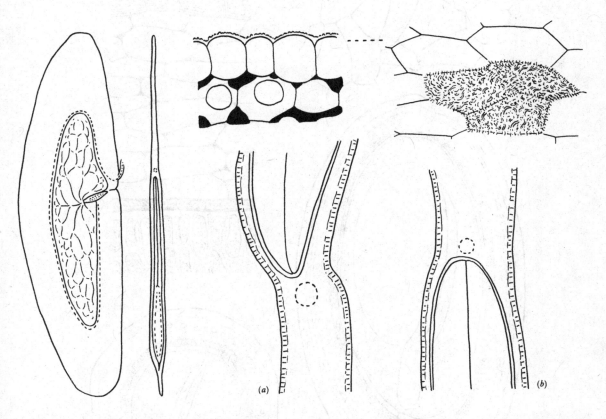

90 *Lophopetalum sp.* (Penang). Seed in median l.s.,
× 1; in t.s., × 4; (*a*) raphe in t.s., (*b*) antiraphe in t.s.,
with v.b. and exotestal cells, × 25. Outer part of the testa
from the side of the seed, × 225.

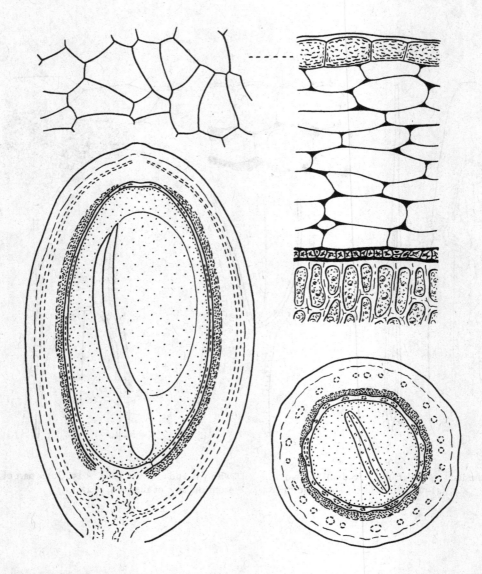

91   *Microtropis platyphylla*. Fruit in l.s. and t.s., with tanniniferous exocarp, woody endocarp-valves, and vascular testa, × 5. Seed-coats in t.s., with the tegmen reduced to a layer of small fibres, the exotesta with tannin-cells, × 225.

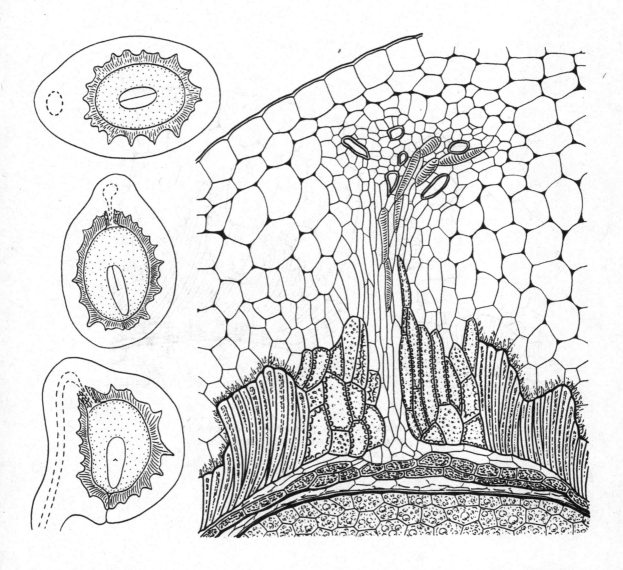

92 *Perrotetia alpestris* ssp. *philippinensis*. Seed in t.s., median and transmedian l.s., ×25. Chalaza in trans-median l.s., with tannin-cells in the endotegmen (dark contents), ×225.

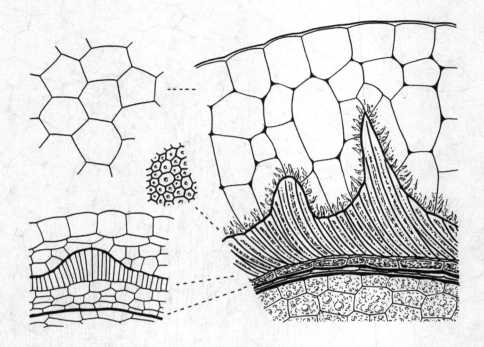

93 *Perrotetia alpestris* ssp. *philippinensis*. Seed-coats of
the young seed with nucellus and of the mature seed with
endosperm in t.s., ×225.

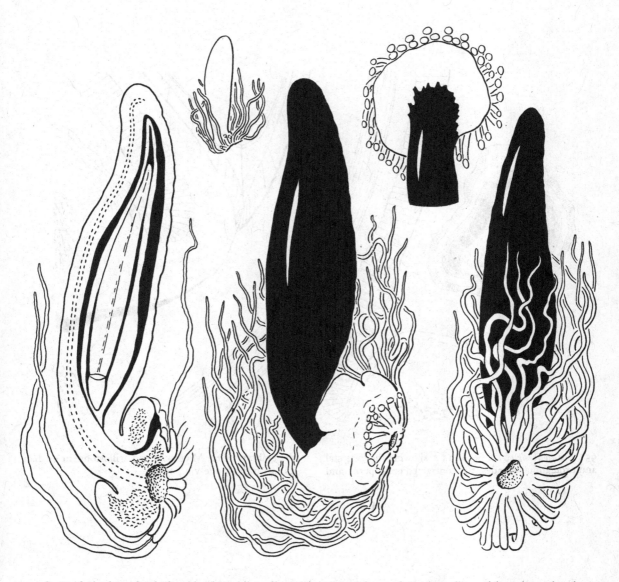

94 *Sarawakodendron*. Seeds in side-view, hilar view, and l.s., with the micropylar cushion-like aril, and the tentaculiferous funicular aril; seed-base in raphe-view, and an unfertilised seed; × 5.

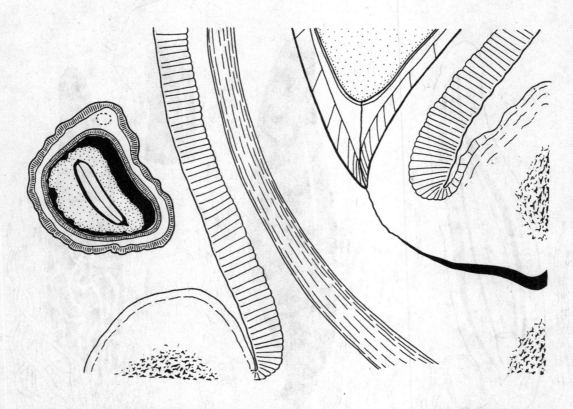

95 *Sarawakodendron*. Seed in t.s. showing the exotestal and exotegmic palisades, the contracted endosperm, and the embryo, × 10. Micropyle and arillostome in median l.s., with the raphe v.b., × 25.

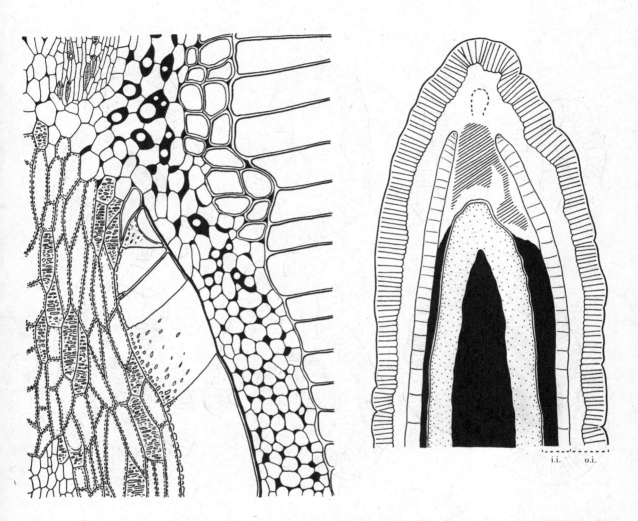

96   *Sarawakodendron*. Chalaza in transmedian l.s., with the lignified tracheidal tissue (hatched), ×25. Microscopic detail of one side of the chalaza, showing the large fibre-cells of the exotegmen and the bases of the palisade-cells of the exotesta, ×115.

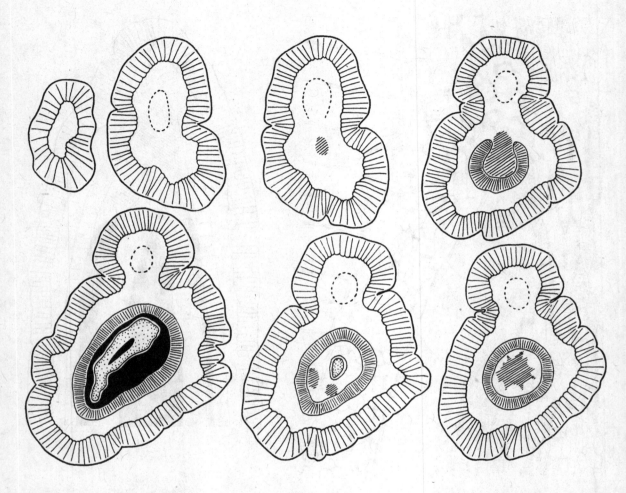

97 *Sarawakodendron.* Chalaza in a series of transverse sections from the apex of the seed to the beginning of the endosperm; exotesta and exotegmen striated, the tracheidal cone of the chalaza hatched; ×25.

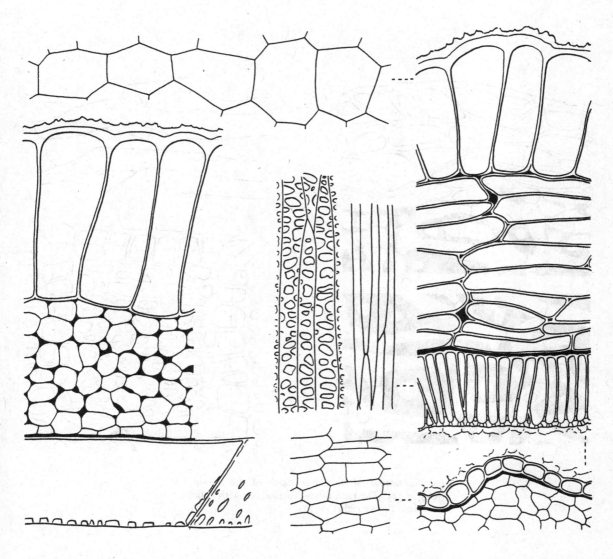

98 *Sarawakodendron*. Seed-coats in l.s. (left) and t.s. (right), the tegmen (right) with disrupted mesophyll, ×225; exotegmic fibres in outer surface-view (left centre) and inner surface-view, ×400.

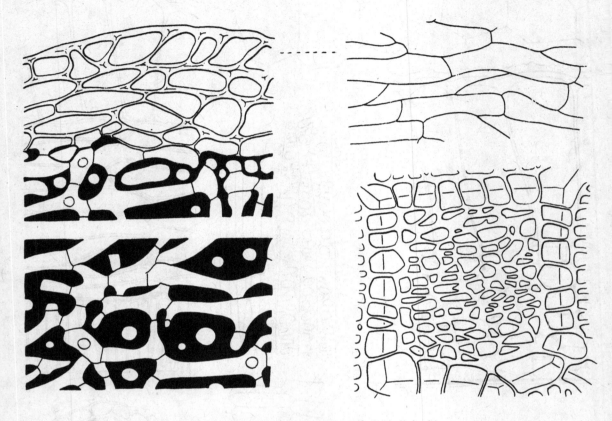

99 *Sarawakodendron*. Outer tissue of the aril in l.s. (left), ×225. The facet of an exotestal cell with cuticular reticulum (lower right), ×400.

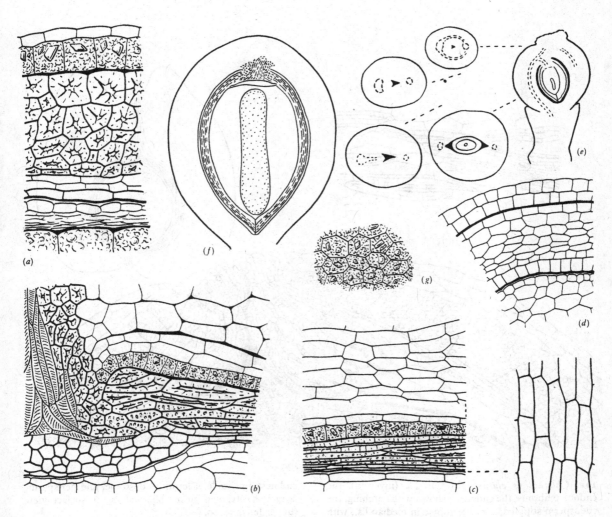

100 *Ascarina maheshwarii*. (*a*) Seed-coat in t.s., the testa 2 cells thick, the tegmen fibrous except for 3 inner cell-layers, the nucellus mostly crushed by the endosperm, × 500. (*b*) Part of the chalaza, the sclerotic sheath continuous with the fibrous exotegmen, the hypostase with the endotegmen, × 225. (*c*) Seed-coat in l.s. with the outer layer of the nucellus, with exotestal and endo-tegmic facets, × 225. (*d*) Ovule-wall in t.s., × 225. (*e*) Ovary in l.s., shortly after pollination, with t.s. at 4 levels, × 10. (*f*) Fully grown but immature berry in l.s., the seed-coat matured, the nucellus partly replaced by the incipient endosperm, × 25. (*g*) Endotestal facets, × 225.

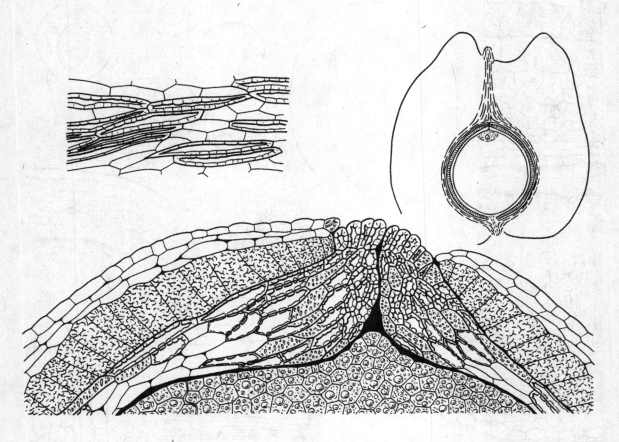

101 *Chloranthus elatior*. Fruit in l.s. (inverted), the endocarp fibrous, the endotesta shown with hatching, the endosperm stippled, × 8. Micropyle in median l.s., with endotestal palisade, sclerotic endostome, and perisperm, × 120. Exotegmen near the chalaza in-surface-view (upper left), × 100.

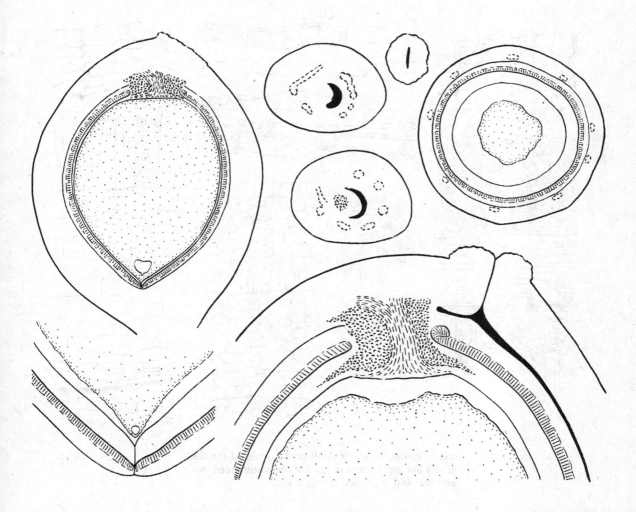

102 *Chloranthus glaber*. Mature seed in l.s. in the berry, ×10; chalazal end with sclerotic tissue at the entrance of the v.b., with crushed remains of the nucellus, ×25.

Immature seed in t.s. in the berry, with t.s. at the placenta and distally from the stigma, ×10. Micropyle of the mature seed in l.s., ×10.

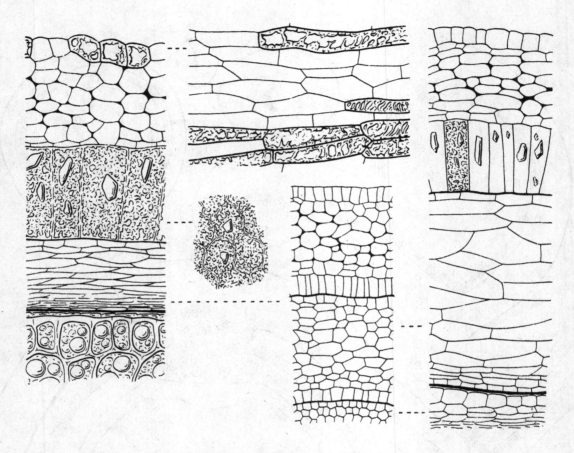

103 *Chloranthus glaber*. Wall of the ovule, of the fully grown but immature seed, and of the mature seed, with exotestal and endotestal facets, ×225.

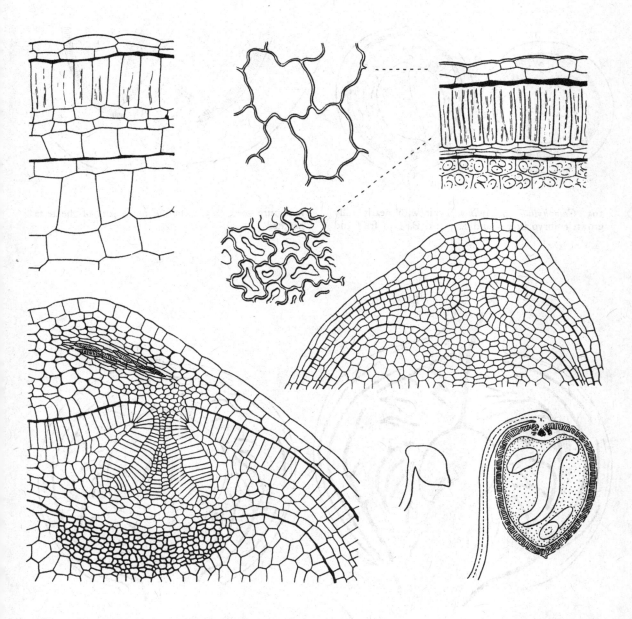

104 *Cistus corbariensis*. Ovule and seed in l.s., × 25. Chalaza of ovule and of fully grown but immature seed in transmedian l.s., × 115. Seed-coats of immature and mature seeds in t.s., × 225; facets, × 500.

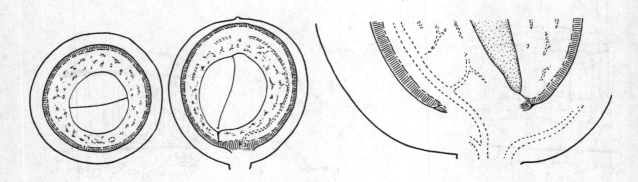

105 *Calophyllum inophyllum*. Fruit with nearly fully grown embryo in t.s. and l.s., × 1. Base of fruit and immature seed, × 5. Sclerotic outer part of the testa hatched.

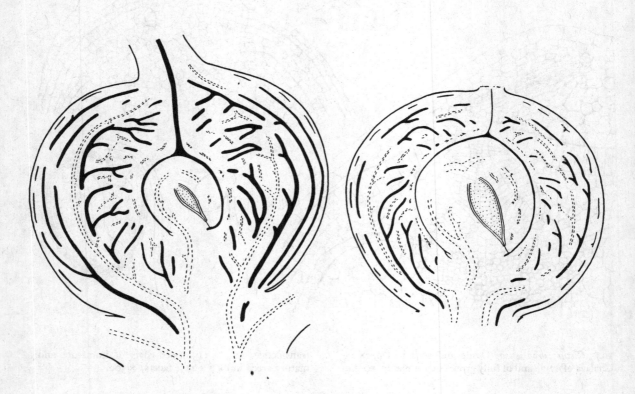

106 *Calophyllum inophyllum*. Ovary at anthesis in l.s. (left), × 25. Young fruit in l.s. before lignification of the testa, × 10. Gum-canals and loculus shown in black.

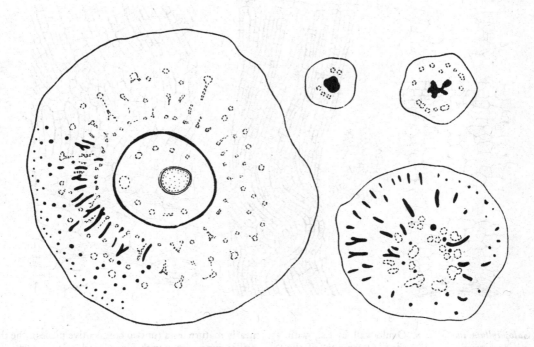

107 *Calophyllum inophyllum*. Ovary in t.s. at various levels from the style to the base, ×25. Gum-canals and loculus shown in black.

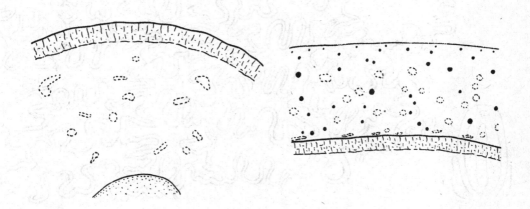

108 *Calophyllum inophyllum*. Nearly mature testa in l.s. (left) with sclerotic outer layer, ×8. Pericarp in t.s. (right) at the same stage, with the woody layer of the testa as the false endocarp, ×8. Gum-canals shown in black.

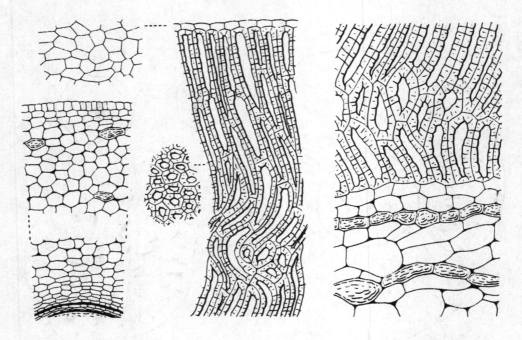

109 *Calophyllum inophyllum*. Ovule-wall in t.s., with thick o.i. and very thin i.i., ×225. Outer part of the nearly mature testa (in two consecutive pieces), the thin-walled inner tissue with files of dark cells, ×225.

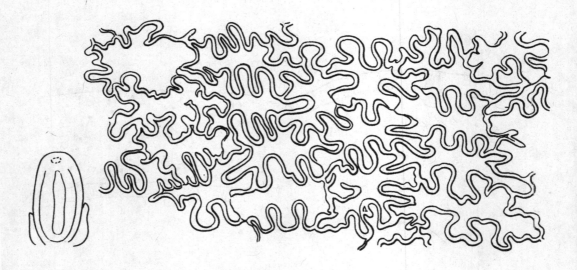

110 *Clusia rosea*. Ovule of flower-bud in transmedian l.s., with incipient aril, ×50. Outer epidermis of the tegmen in surface-view, ×115.

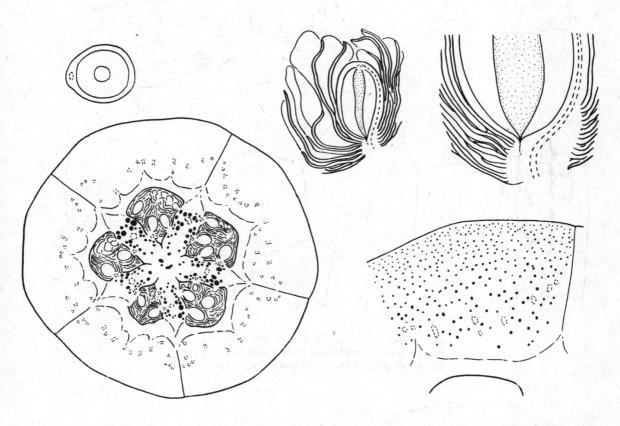

111 *Clusia sellowiana*. Immature fruit in t.s., showing the parietal placentation, the latex-tubes (in black) of the core, v.b. in the mesocarp, × 5. Pericarp in t.s. with the latex-tubes shown in black, × 10. Immature seed with aril-lobes in l.s., and a seed without aril-lobes in t.s., × 10. Base of the immature seed, showing the multiple origin of the aril, × 25.

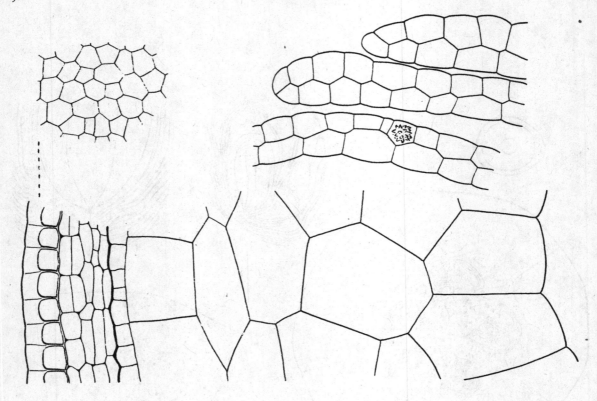

112 *Clusia sellowiana*. Seed-coats in t.s. from an immature seed; tegmen with small-celled o.e. and large mesophyll-cells, × 225. Aril-lobes in l.s., × 225.

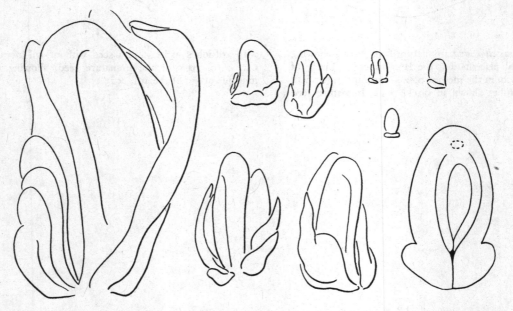

113 *Clusia sp.* Immature seed with multiple aril, from the raphe-side, × 25. Ovule and developing seeds, × 25.

Ovule of the flower-bud in transmedian l.s., with incipient aril, × 115.

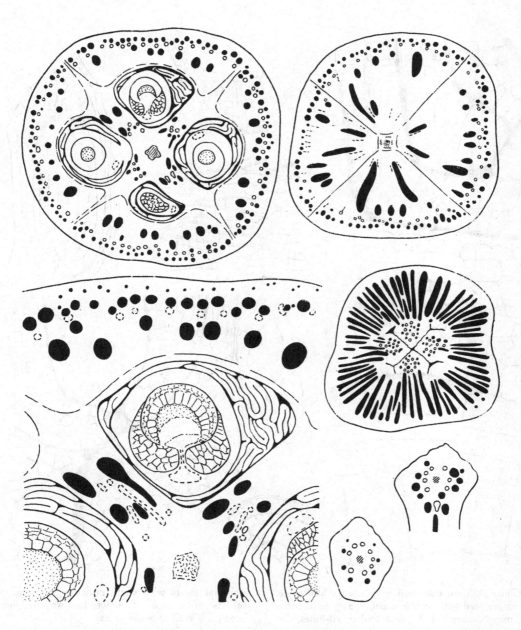

114 *Clusia sp.*, fruit-structure. Fruit in t.s. at various levels from the stigma (lower right) to the seeds, × 10.

Part of the fruit in t.s., × 25. Latex-tubes, loculi, and stylar-canal shown in black.

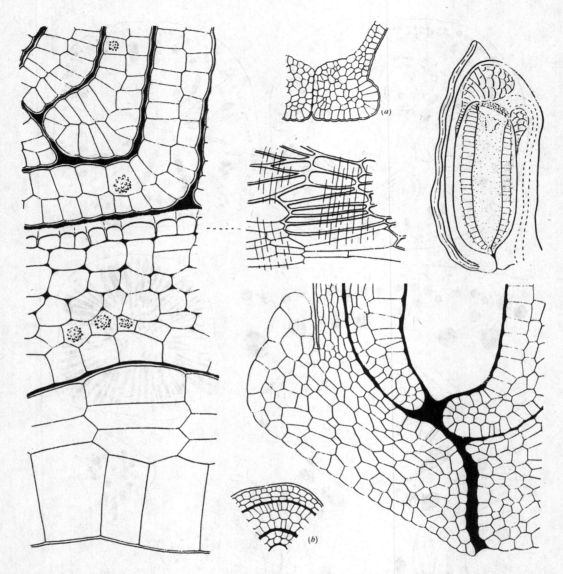

115 *Clusia sp.* Immature seed in l.s., ×25. Micropyle of a younger seed with incipient aril, ×225. Seed-coats of the immature seed in t.s., with folded aril-lobes, the exotestal facets with slight cuticular striation, ×225. (*a*) Base of the ovule of a flower-bud with incipient aril, ×225. (*b*) Wall of ovule, ×225.

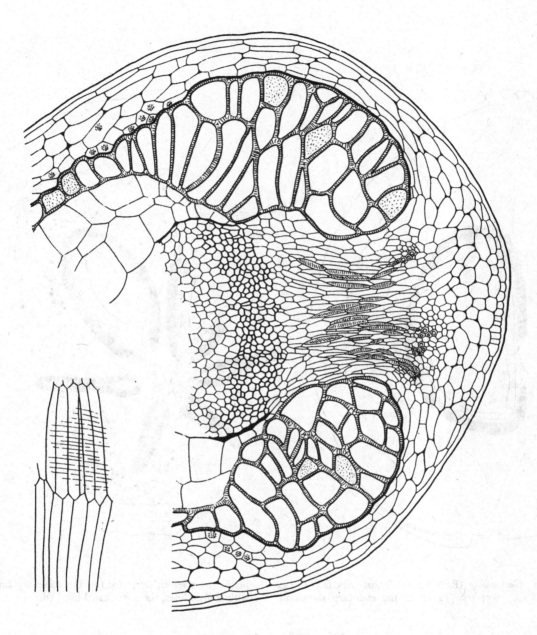

116 *Clusia sp*. Chalaza in t.s., with sclerotic exotegmen
and brown hypostase, × 115. Exotestal cells over the
chalaza, × 225.

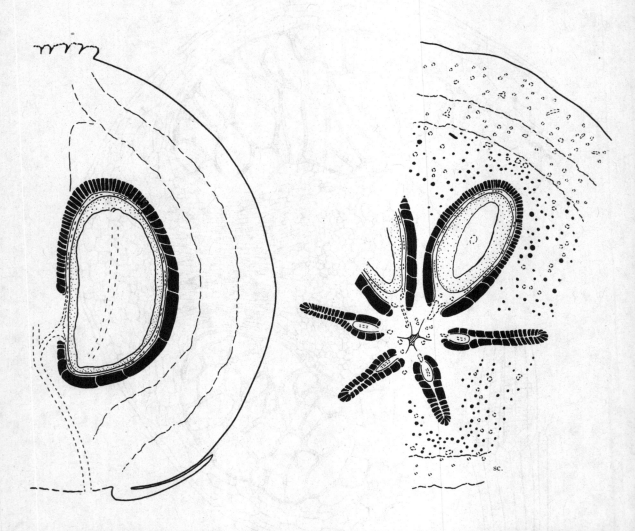

117 *Garcinia sp.* (RSNB 4085). Nearly ripe fruit in l.s. and t.s., ×5; resin-canals of the endocarp shown in black, the sclerotic layer (sc.) of the mesocarp limited with broken lines, the stylar canal hatched.

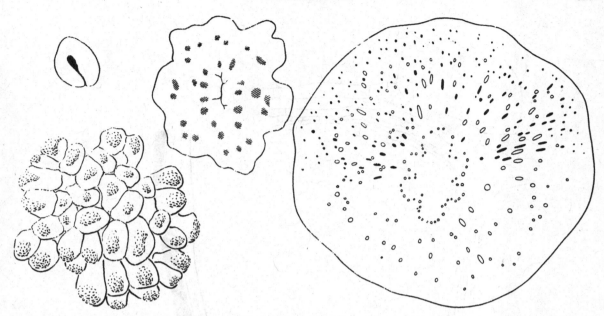

118 *Garcinia sp.* (RSNB 4085). Cluster of 38 stigmata, and t.s. of the fruit at the base of the stigmata with *c.* 30 stylar canals, × 10; stigmatic cleft in t.s., × 25. Fruit in t.s. distal to the loculi, with the resin-canals shown in black, × 8.

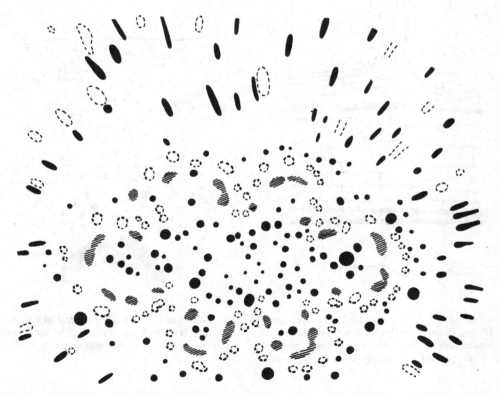

119 *Garcinia sp.* (RSNB 4085). The centre of the fruit in t.s. just distal to the loculi, showing 27 stylar canals (hatched) with attendant v.b., conforming to the six loculi; resin-canals shown in black.

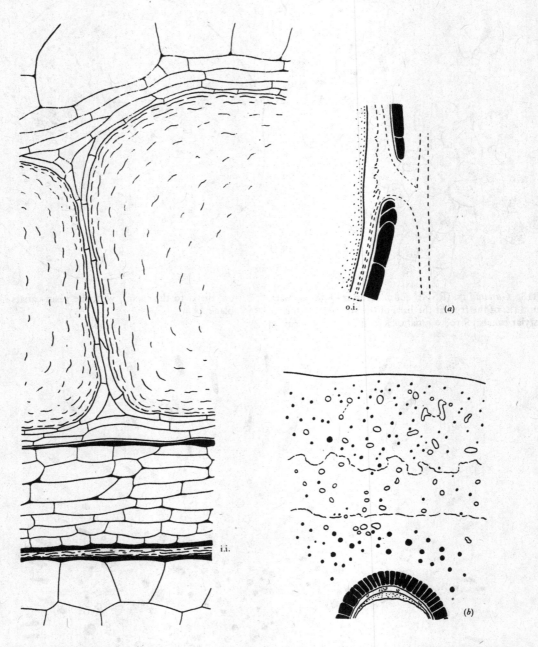

120 *Garcinia sp.* (RSNB 4085). Endocarp and seed-coats (left) in t.s., showing the large resin-canals of the endocarp surrounding the seed, the collapsed tegmen, and the outer part of the watery endosperm, × 225.

(*a*) Hilum in l.s., × 8. (*b*) Pericarp in t.s., showing the sclerotic layer with v.b. (delimited by wavy broken lines), v.b. as open circles, and resin-canals in black, × 10.

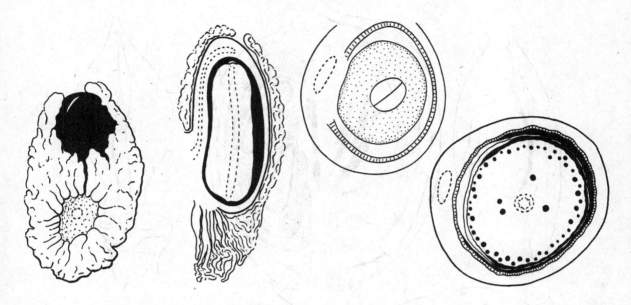

121 *Havetiopsis flexilis*. Seeds in hilar view and l.s., with multiple aril, × 10. Seeds in t.s. across the coty- ledons and across the hypocotyl (latex-tubes in black), with the exotegmen striated, × 25.

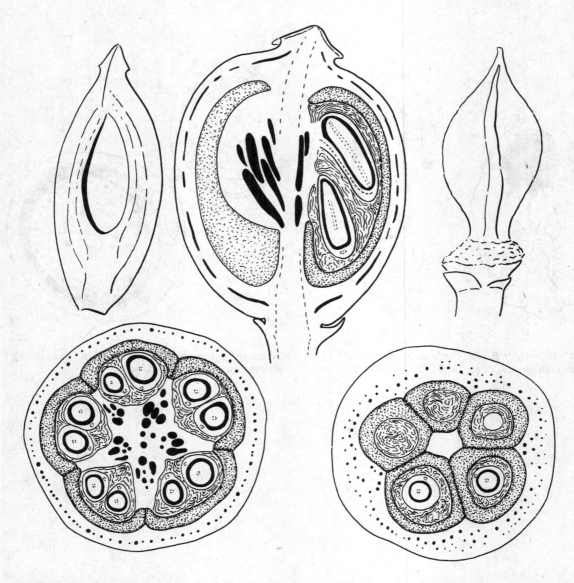

122 *Havetiopsis flexilis*. Mature fruit in l.s. and t.s. across the middle (left) and near the base (right), with the bony endocarp speckled, the aril-lobes as broken lines, the latex-tubes and loculi in black, × 5. A valve and the columella of the dehisced fruit, × 5.

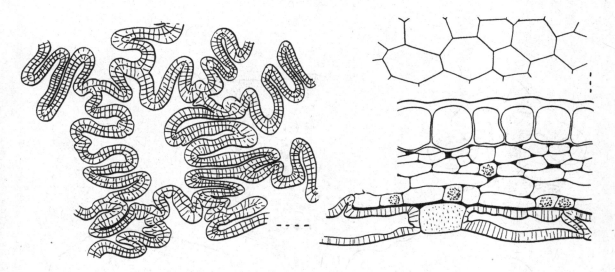

123 *Havetiopsis flexilis*. Testa in t.s., with the facets of the exotesta and exotegmen (lignified stellate cells), × 225.

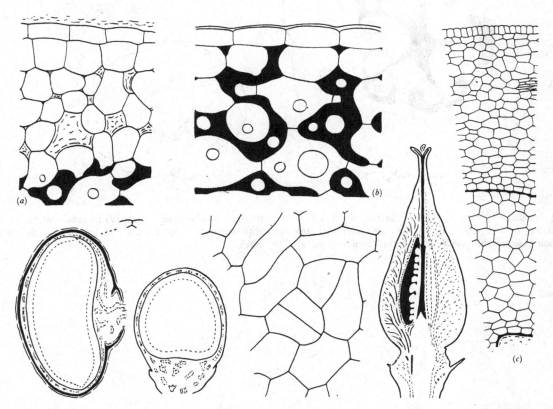

124 *Pentadesma butyracea*. Seeds in l.s. and t.s., with hypocotylar embryo, vascular testa, and discoid aril, × 1. Ovary in l.s., showing v.b. (right half) and latex-tubes (left half), × 1. (*a*) Aril in t.s.; (*b*) outer part of the testa; (*c*) wall of the ovule, × 225.

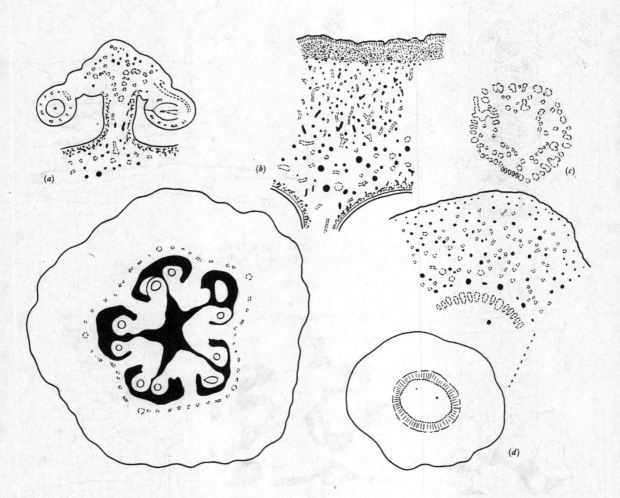

125 *Pentadesma butyracea*. Ovary in t.s., × 5. (*a*) Placenta in t.s., × 10. (*b*) Pericarp in t.s., with sclerotic outer region beneath the periderm, × 5. (*c*) Vascular tissue at the base of the ovary, × 10. (*d*) Peduncle in t.s. (× 5), with cortex (× 10). Loculi and latex-tubes shown in black.

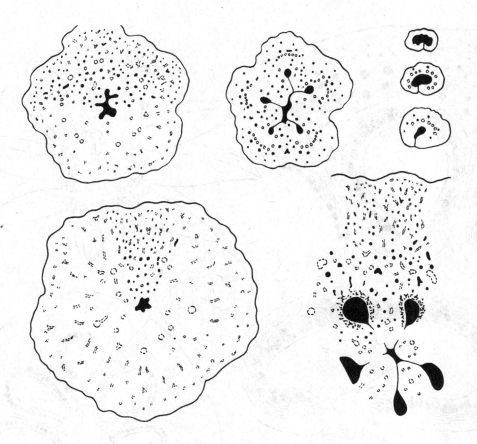

126 *Pentadesma butyracea.* Stigma, style, and apex of the ovary in t.s. at various levels, × 10. Latex-tubes and loculi shown in black.

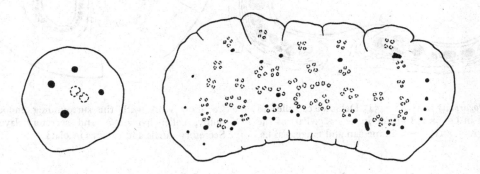

127 *Pentadesma butyracea.* Staminal phalanx in t.s. (right), × 10; base of a filament with the 4 v.b. joined in pairs (left), × 25. Latex-tubes in black.

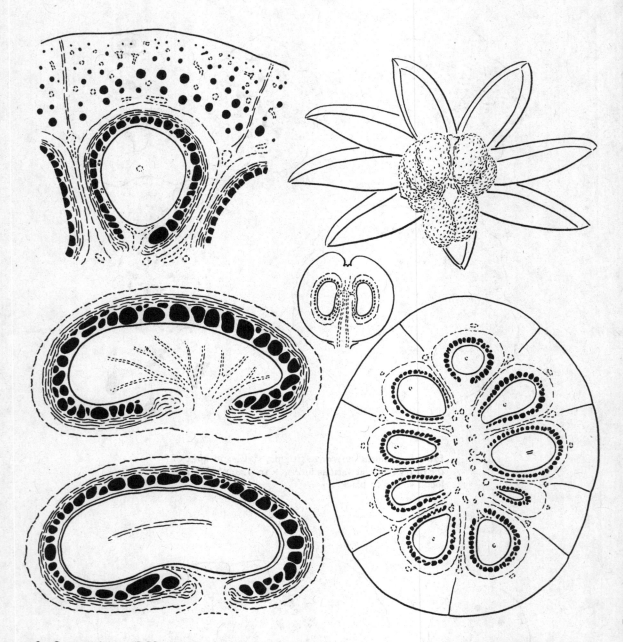

128 *Septogarcinia sp.* (RSS 2454). Dehisced capsule and indehisced in l.s., × 2. Capsule in t.s., × 5. Part of the capsule in t.s., × 10. Seeds in median and tangential l.s. (to show v.b.), with the surrounding endocarp with pulpy, collenchymatous, and secretory layers, × 10. Secretory canals and cavities in black.

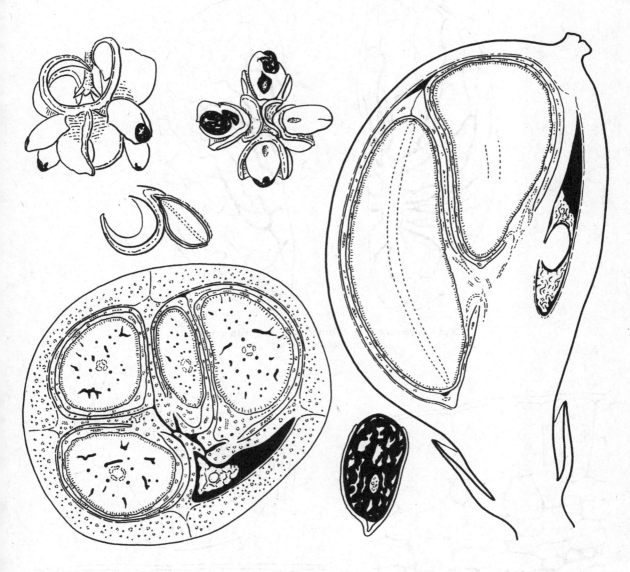

129 *Tovomitopsis sp.* (Corner 255). Dehisced fruits, × 1. Seed in l.s., attached by the aril to the fruit-valve, × 1½. Mottled seed with the aril cut off, × 3. Fruit in l.s., with two seeds (hypocotylar embryo) in one loculus, the other with an abortive seed, × 5. Fruit in t.s., with two loculi 2-seeded, one 1-seeded, one with an abortive seed, × 5. Latex-tubes and loculi shown in black.

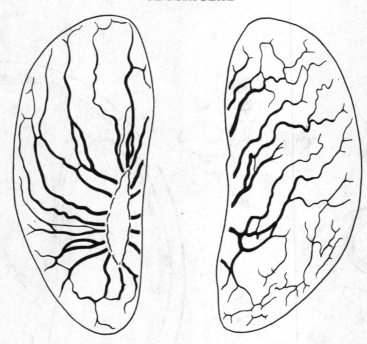

130 *Tovomitopsis sp.* (Corner 255). Aril with venation
on the two sides, × 5.

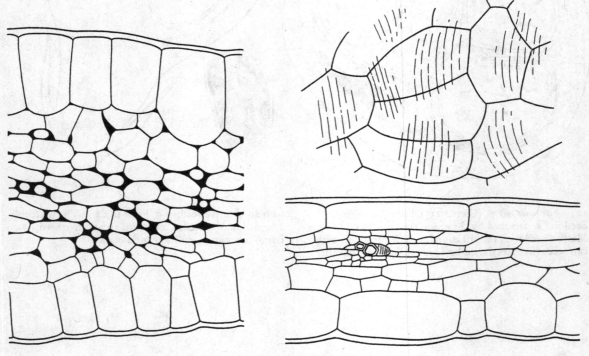

131 *Tovomitopsis sp.* (Corner 255). Aril in t.s. at the
thicker and thinner parts, with the epidermal facets,
× 225.

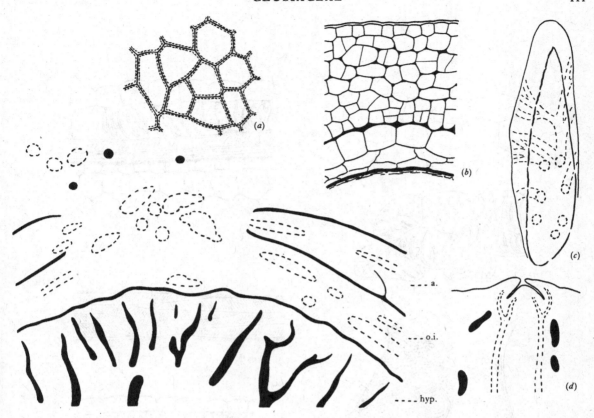

132 *Tovomitopsis sp.* (Corner 255). Hilum in t.s. with aril (a.), testa (o.i.), and embryo (hyp.), the latex-tubes shown in black, × 25. (*a*) Facets of the exotesta, × 225. (*b*) Wall of the young seed in t.s., with crushed nucellus, tegmen becoming crushed, and proliferating testa, × 225. (*c*) Young seed in hilar view with v.b., × 10. (*d*) Apex of the embryo with cotyledons, × 25.

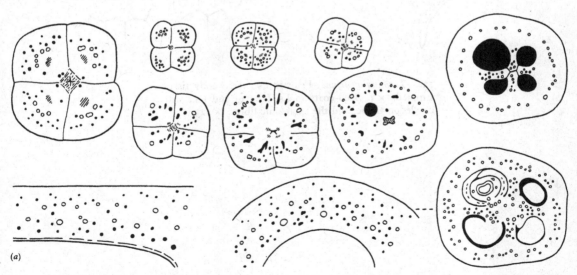

133 *Tovomitopsis sp.* (Corner 255). Young fruit in t.s. at various levels from the stigma to the placentas, × 10; wall of the young fruit, × 25: (*a*) part of the mature pericarp, in t.s., × 10. Latex-tubes and loculi shown in black, v.b. as open circles, stylar canal hatched.

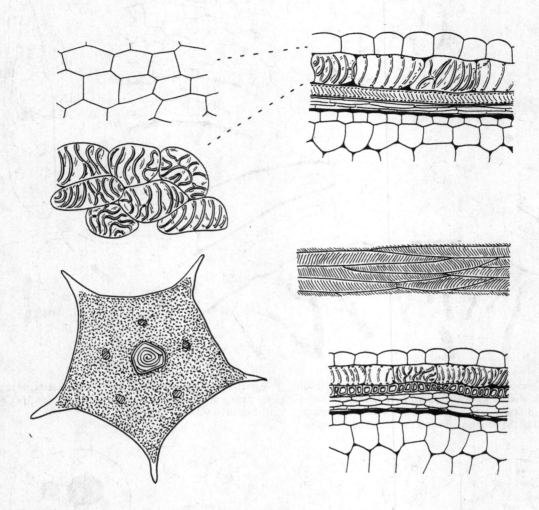

134 *Terminalia arjuna*. Fruit in t.s., the woody tissue
stippled, × 3. Seed-coats in l.s. (above) and t.s., with
cotyledon-tissue, and the exotegmic fibres, × 225.

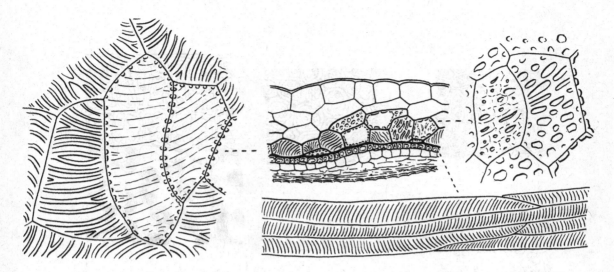

135 *Terminalia parviflora*. Seed-coat in t.s. (× 225) with
tangential views (× 500) of the mesotestal sclerotic cells,
the endotestal tracheids, and the exotegmic tracheids.

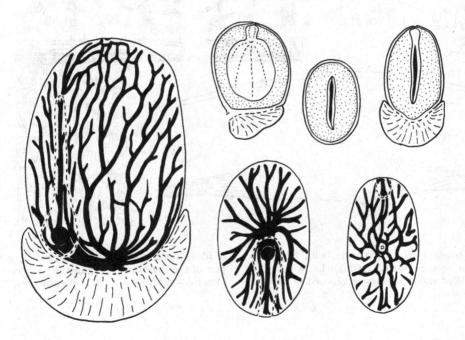

136 *Cnestis palala*. Seeds in median and transmediam
l.s. with the aril, and in t.s., × 1. Vascular supply of the
seed in side-view, chalazal (basal) view, and micropylar
(apical) view, with the area of the hilum shown by broken
lines, × 3.

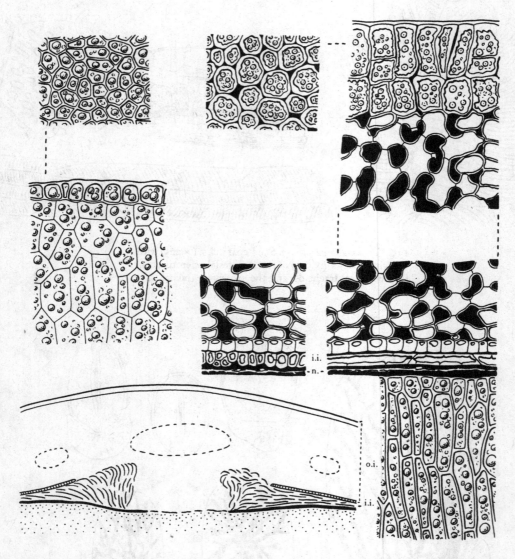

137 *Cnestis palala*. Chalaza in section, with v.b. and fibrous tegmen, × 30. Seed-coats (right) in l.s., with thick testa, thick-walled endotesta, tegmen reduced to a layer of fibres, nucellar cuticle, and endosperm, × 225; inner part of the testa and the tegmen in t.s. (centre), × 225. Outer tissue of the aril (left) with oily cells, × 225.

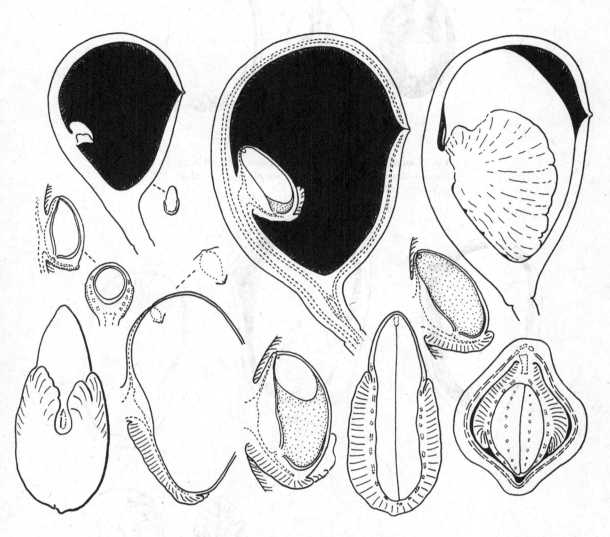

138 *Connarus grandis*. Developing follicles in l.s.,
mature follicle in l.s. and t.s., and mature seed, × 1.
Young seeds, × 2.

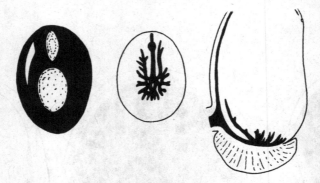

139 *Connarus monocarpus*. Hilar view of the seed with the aril-attachment, × 2. Vascular supply of the seed in side-view and chalazal view.

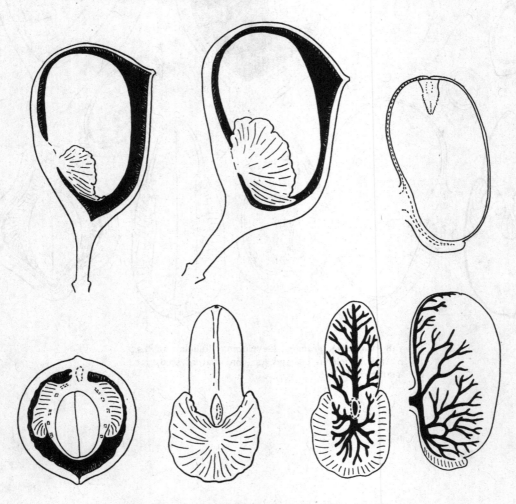

140 *Connarus semidecandrus*. Follicles and seed in l.s. and t.s., the seed in hilar view, and the vascular supply of the seed in hilar and side-view, × 3.

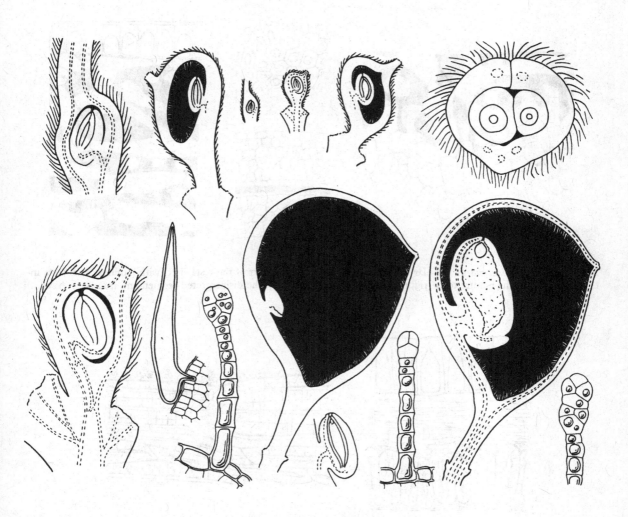

141 *Connarus semidecandrus*. Ovary and young fruits in l.s., ×7; enlargements (left) ×30. Follicles (fully grown) with developing seeds, ×3; enlargement of the smaller seed in l.s., ×6. Ovary in t.s. (upper right), ×50. Endocarp-hairs, ×225.

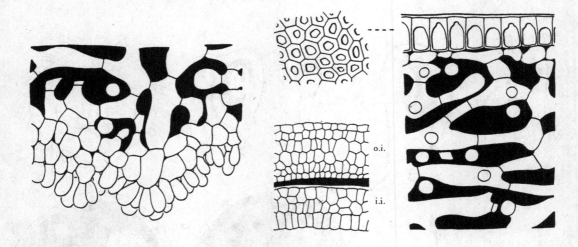

o.i.

i.i.

142 *Connarus semidecandrus.* Ovular wall (centre) in t.s.; outer part of the mature testa (right) in t.s., and the inner part of the testa (left) with papillose endotesta invading the disintegrated tegmen and nucellus; × 225.

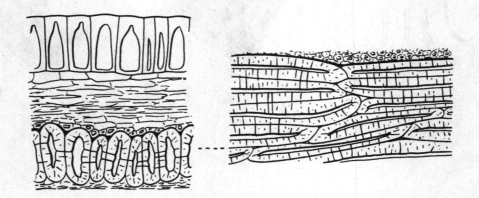

143 *Connarus villosus.* Mature seed-coat in t.s. with the exotestal palisade and the endotestal crystal-cells, with the fibres of the exotegmen in t.s. and l.s., × 225.

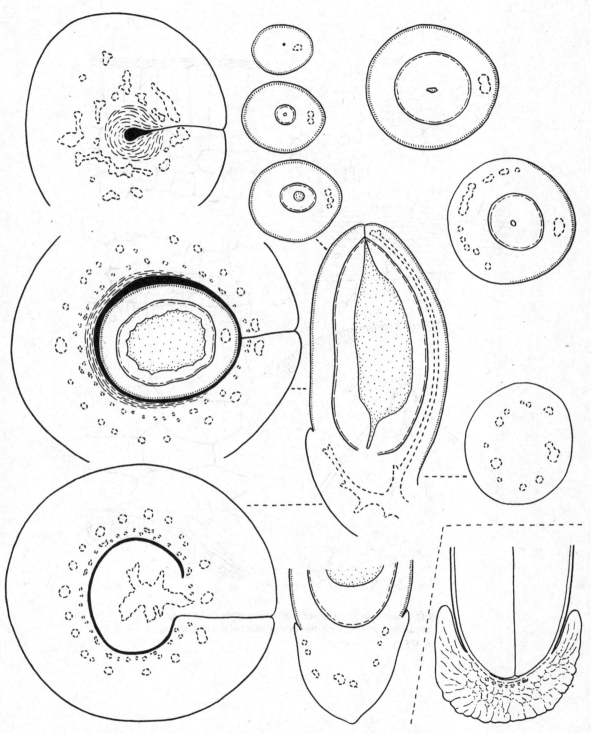

144   *Connarus sp.* (Corner 256). Immature seed in l.s.,   × 10. Inset, the base of the seed of *Connarus salomonensis*,
t.s., with the base of the seed in transmedian l.s. (showing   × 2.
the slight aril-rim), and 3 sections of the follicle in t.s.,

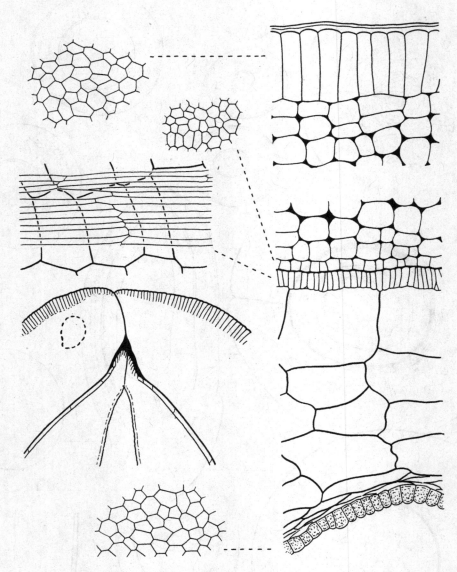

145 *Connarus sp.* (Corner 256). Micropyle in l.s., × 50. Seed-coats in t.s., with the exotegmen narrowly fibrous, × 225.

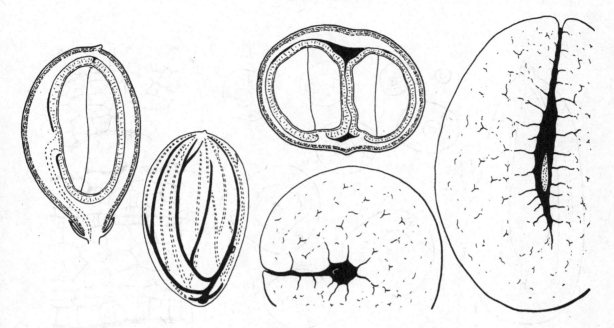

146 *Jollydora duparquetiana*. Fruit in l.s. (1-seeded) and in t.s. (2-seeded) with the sclerenchymatous mesocarp stippled and the sarcotesta striate, × 2. Vascular supply of the seed, × 3. Seeds in hilar and micropylar views, with the black preraphe devoid of sarcotesta, × 5.

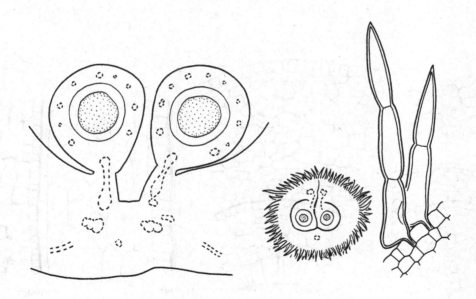

147 *Jollydora duparquetiana*. Ovary in t.s., × 25; placenta with ovules in t.s., × 100. Hairs on the ovary, × 225.

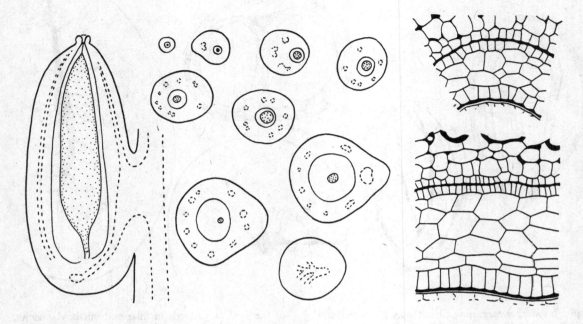

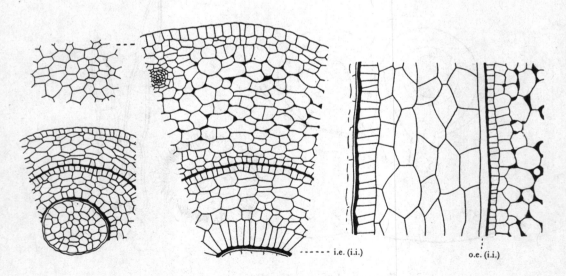

148 *Jollydora duparquetiana*. Young seed in l.s. and in t.s. at various levels, × 10. Tegmen of this seed in t.s., with the inner part of the testa, near the micropyle (above) and in the body of the seed, × 225.

i.e. (i.i.)

o.e. (i.i.)

149 *Jollydora duparquetiana*. Wall of ovule and of young seed in t.s.; tegmen of an older seed in l.s.; × 225.

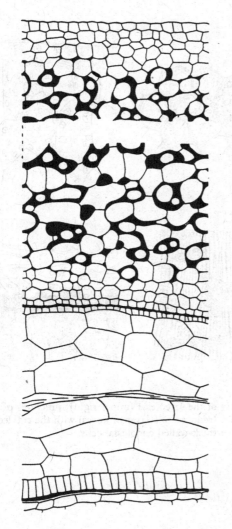

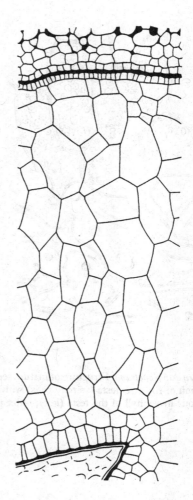

150 *Jollydora duparquetiana*. Seed-coats of a fully grown, but not yet sarcotestal, seed in t.s. (left), with the central cells of the tegmen collapsing, ×225. Tegmen in t.s. near the hypostase (right), ×225.

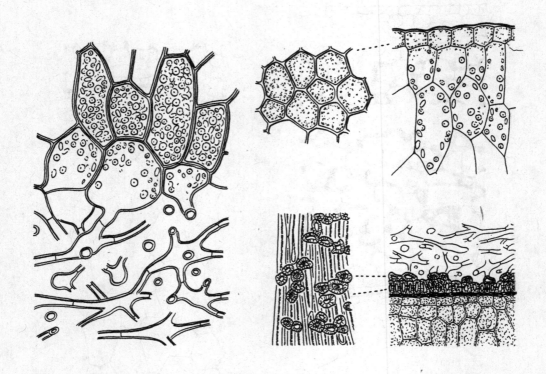

151　*Jollydora duparquetiana*. Structure of mature seeds, ×225; junction of the sarcotesta (starchy cells) with the aerenchymatous mesophyll of the testa (left); outer part of the sarcotesta (upper right); junctions of the tegmen (reduced to a layer of fibres) with the cotyledon and the thick-walled endotestal cells.

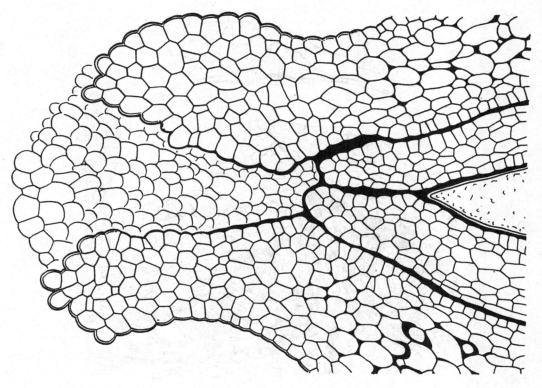

152 *Jollydora duparquetiana.* Micropyle of an immature
seed in l.s. × 225.

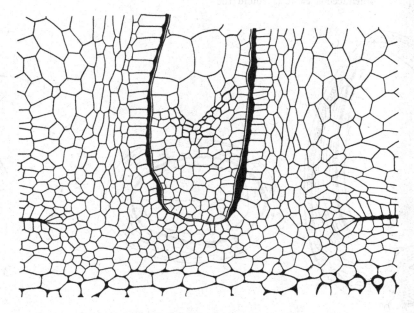

153 *Jollydora duparquetiana.* Nucellar base of the immature seed in l.s., showing the origin of the tegmen and
testa, × 225.

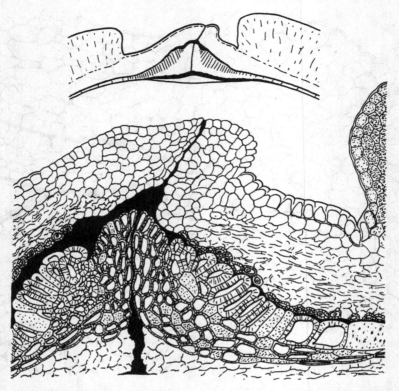

154 *Jollydora duparquetiana*. Micropyle of the mature sarcotestal seed in l.s., ×25; details of the central part, showing the typical Connaraceous exotesta round the exostome, with the sarcotesta on the right, the endostome wholly sclerotic, ×125.

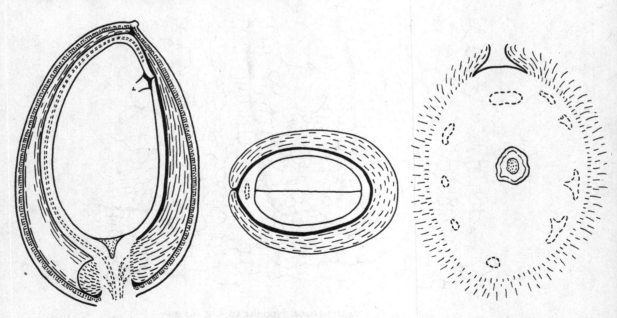

155 *Rourea minor*. Fruit in l.s., with the endocarp striated and the aril with broken lines, ×2. Seed in t.s., with v.b. in the preraphe, ×2. Seed-base in t.s. with the aril attached all round except at the preraphe, ×10.

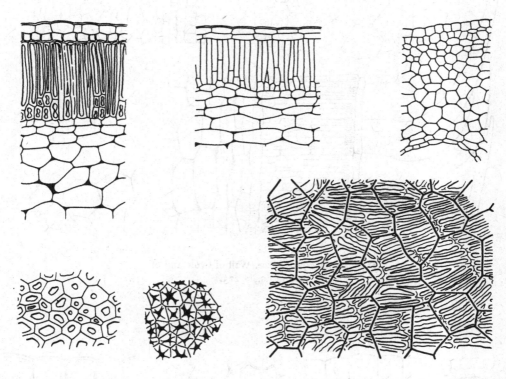

156 *Calystegia sepia*. Upper row; wall of ovule (right), of young seed (centre) and of mature seed (left), × 225. Lower row; facets of the large outer epidermal cells and of the small hypodermal cells (right); palisade-cells with their outer facets and in t.s. about the middle region; × 400.

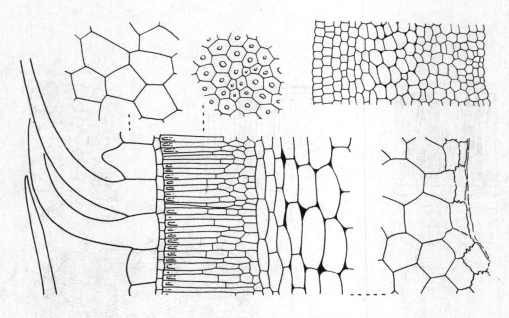

157   *Ipomaea pes-caprae*. Wall of ovule and of
seed in t.s., ×225.

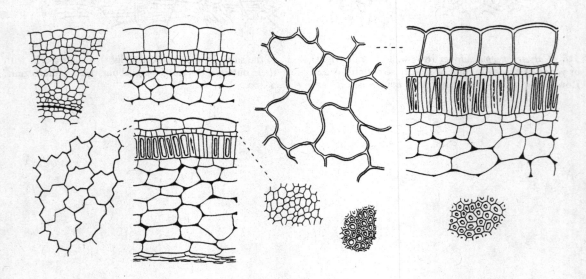

158   *Jacquemontia* (upper row); wall of ovule, of young
seed, and of mature seed in t.s., ×225. *Evolvulus* (lower
row); wall of nearly mature seed in t.s., ×225.

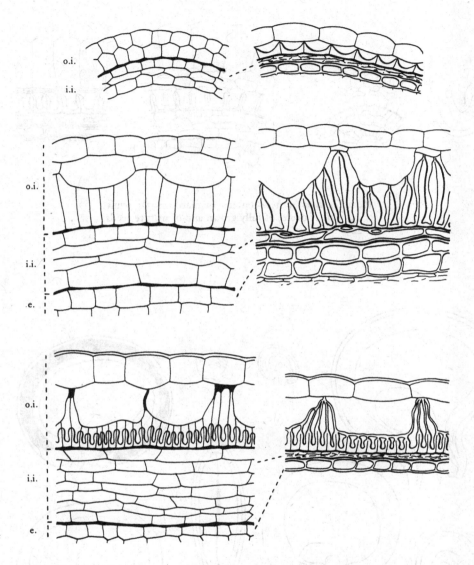

159 Cruciferous seed-coats in t.s., ×225 (after Guignard 1893). *Capsella bursa-pastoris* (top row), ovule and mature seed-coat. *Thlaspi arvense*, young and mature seed-coats. *Brassica nigra*, (bottom row), fully grown and mature seed-coats.

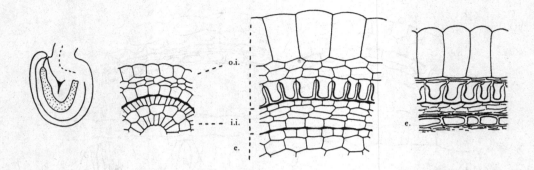

160 *Lepidium sativum* (after Guignard 1893); ovule,
× 25; wall of ovule, of fully grown and of mature seeds,
× 225.

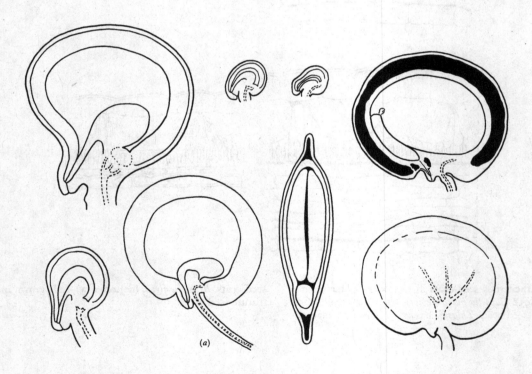

(a)

161 *Lunaria annua*. Ovule and developing seeds in
median l.s., × 25. (a) Nearly fully grown but immature
seed, × 8. Mature seeds in l.s. and in side-view with v.b.,
× 5; in t.s., × 8.

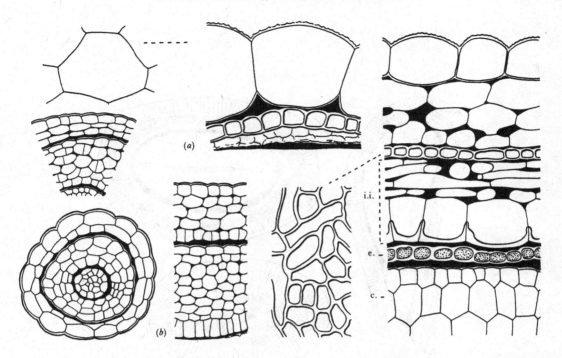

162 *Lunaria annua*. Wall of ovule (left) near the micropyle and near the chalaza; wall of fully grown seed: (*a*) wall of the seed-wing with disrupted tegmen; (*b*) wall of young seed, × 225.

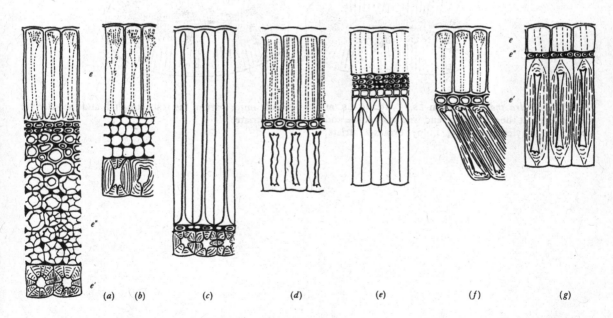

163 Cucurbitaceous seed-coats, showing the outer part of the testa in t.s.: *e*, the seed-epidermis; *e″*, the seed-hypodermis; *e′*, the inner sclerotic layer of short fibres or stellate cells (cuboid in t.s.) or of palisade-cells; after Kratzer (1918). (*a*) *Lagenaria*, (*b*) *Cucurbita pepo*, (*c*) *Cucumis sativa*, (*d*) *Ecballium elaterium*, (*e*) *Luffa aegyptiaca*, (*f*) *Bryonia*, with oblique palisade-cells, (*g*) *Sicyos angulata*.

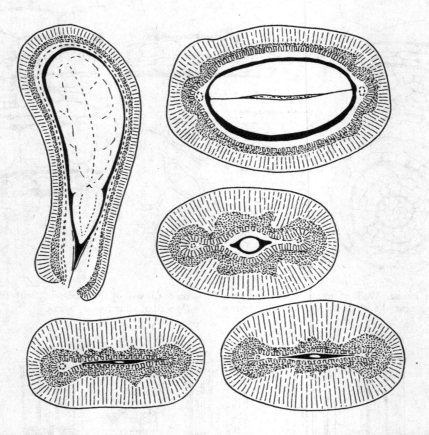

164 *Cephelandra indica*. Seed in l.s., × 10: in t.s. at various levels at the micropylar end with the radicle and across the cotyledons, × 25; pulpy exotesta striate, woody outer part of the testa stippled, and the sclerotic layer striate.

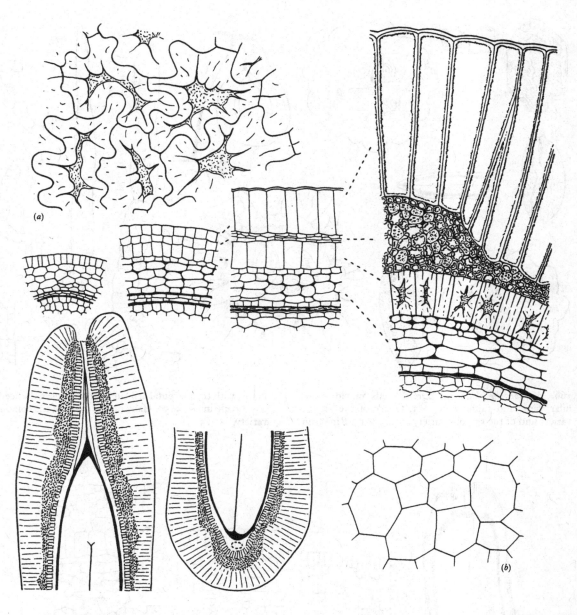

165 *Cephelandra indica*. Wall of ovule, young seeds and mature seeds showing the origin and differentiation of the *e*-, *e″*-, and *e′*-layers, and of the mesophyll of the testa, ×225. (*a*) Facets of the inner sclerotic cells (*e′*-layer), ×500; (*b*) facets of the outer epidermal cells (*e*-layer), ×225. Micropyle and chalaza in transmedian l.s., ×25.

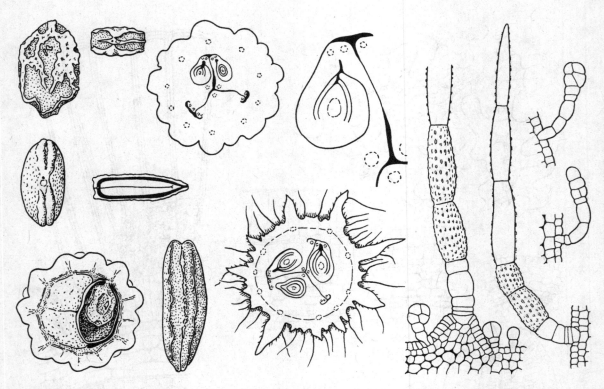

166 *Momordica charantia*. Dried seeds in side-view, hilar view, and raphe-view, × 3. Ovary of the warted variety and of the spinous variety in t.s., × 15. Ripe fruit in t.s., with the placental aril cut open to expose the seed, × 1. Ovule in l.s., × 50. Hairs on the ovary of the spinous variety, × 225.

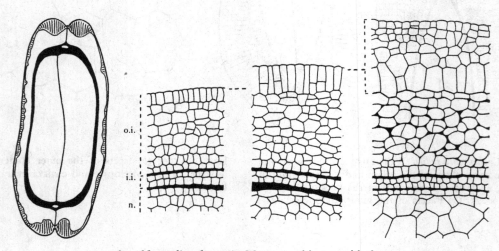

o.i.
i.i.
n.

167 *Momordica charantia*. Mature seed in t.s., with the pulpy epidermis striated, × 7. Wall of the ovule and of developing seeds in t.s., × 225.

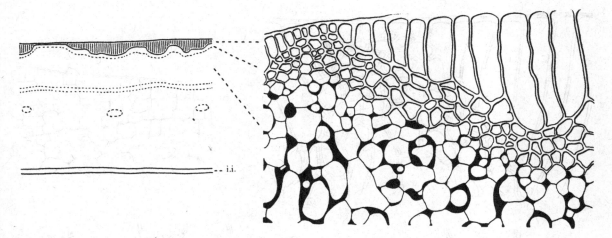

168 *Momordica charantia.* Diagram of the seed-coats in t.s., showing the five layers of the testa, with v.b. in the inner layer, × 30. Outer part of the testa in t.s., × 225.

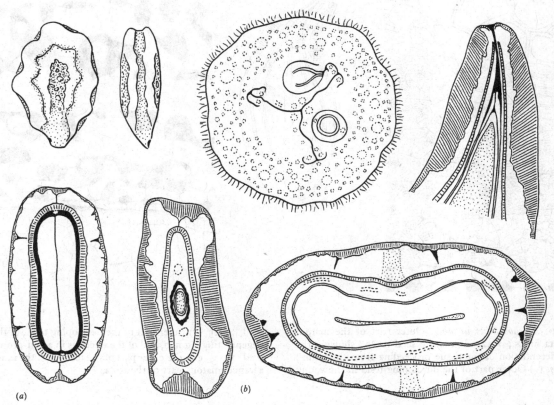

(a)

(b)

169 *Trichosanthes anguina.* Seeds in side-view and raphe-view, with the hilum shown as a black spot, × 2. Ovary in l.s. with v.b. (as procambial strands, except the larger), × 15. Micropylar end of the seed in transmedian l.s., × 7. (a) Mature seed in t.s., × 6; (b) immature seed in t.s. at the micropylar region and the middle region, × 7. The pulpy epidermis and the inner sclerotic layer of the testa striated.

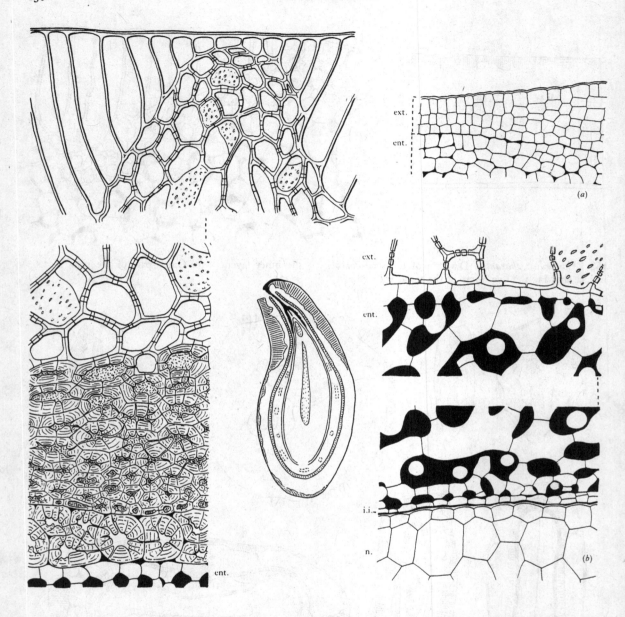

170 *Trichosanthes anguina*. Outer part of the mature testa in t.s., × 225. Immature seed in l.s., showing the differentiation of the tissue proceeding from the hilum, × 7. (a) Outer part of the developing testa in the young seed, × 225. (b) Inner part of the developing testa, of the degenerating tegmen, and of the nucellus in a half-grown seed, × 225. ext. the outer part of the testa; ent. the inner aerenchymatous part of the testa.

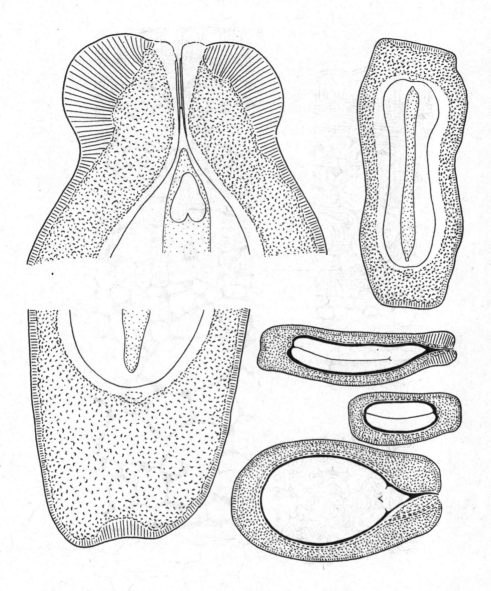

171 *Trichosanthes sp.* (Ceylon). Ripe seeds in median and transmedian l.s. and in t.s., × 5. Chalaza and micropyle of immature seeds in transmedian l.s., and the seed in t.s., × 25. Pulpy epidermal layer striated, the woody part of the testa speckled.

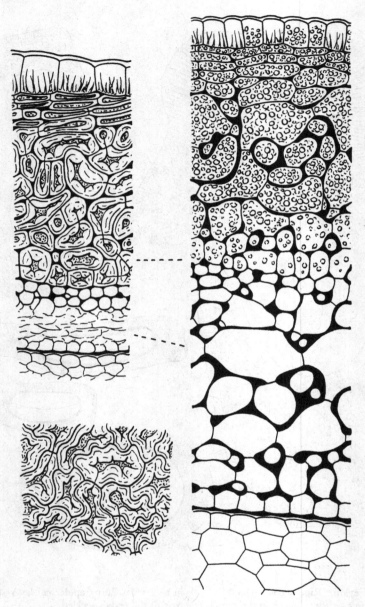

172 *Trichosanthes sp.* (Ceylon). Mature (left) and immature seed-coats in t.s., with nucellar tissue, and the
facets of the inner sclerotic cells of the testa, × 225.

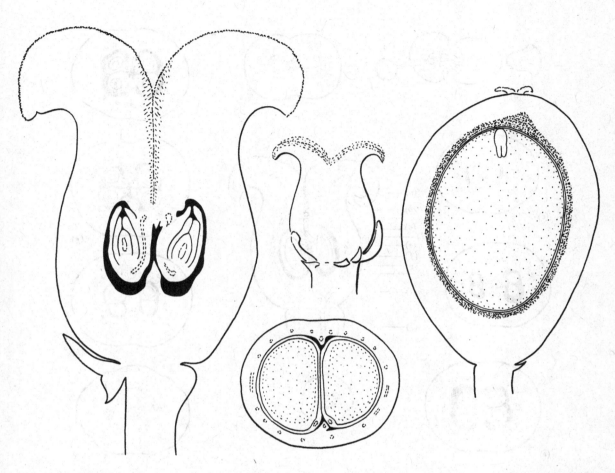

173 *Daphniphyllum borneense*. Female flowers at anthesis, × 10; in l.s., soon after fertilisation, × 25. Ripe fruit in l.s., with woody endocarp, and unripe fruit in t.s. with two immature seeds, × 5.

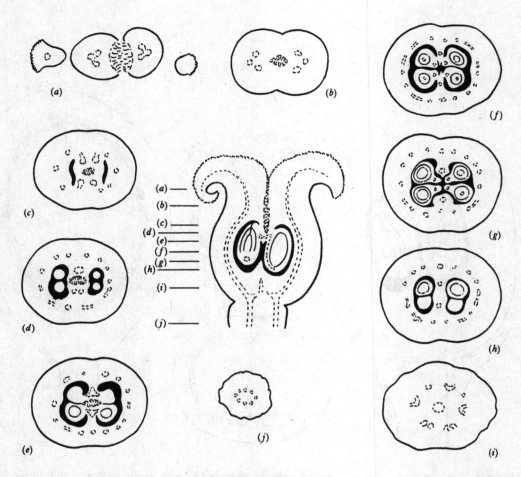

174 *Daphniphyllum borneense.* Female flower in l.s.
and in t.s. at various levels, × 10.

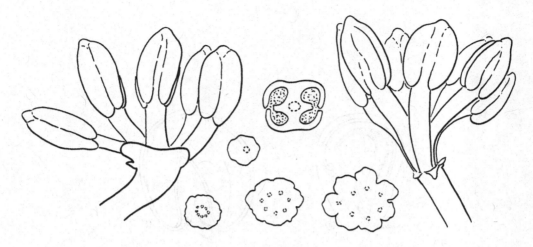

175 *Daphniphyllum borneense*. Male flowers in section
(left), × 8. Anther, filament, receptacle, and pedicel in
t.s., × 10.

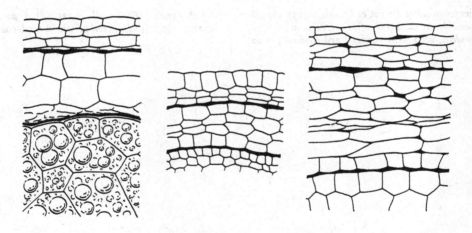

176 *Daphniphyllum borneense*. Wall of ovule (centre), of
immature seed (right) and of mature seed (left) in t.s.,
× 225.

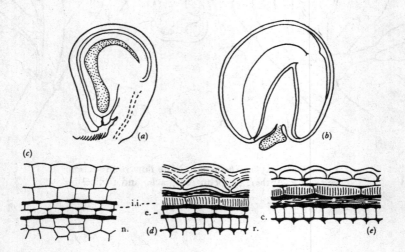

177   Didieraceae (after Perrot et Guérin 1903). (*a*) *Al-*
luardia dumosa, ovule soon after fertilisation, × 30.
(*b*) *A. ascendens*, seed with vestigial aril (dotted), × 12.
(*c*) *A. dumosa*, ovule-wall, × 225. (*d*) *A. ascendens*, seed-
coats, × 225. (*e*) *Didierea mirabilis*, seed-coats, × 225.

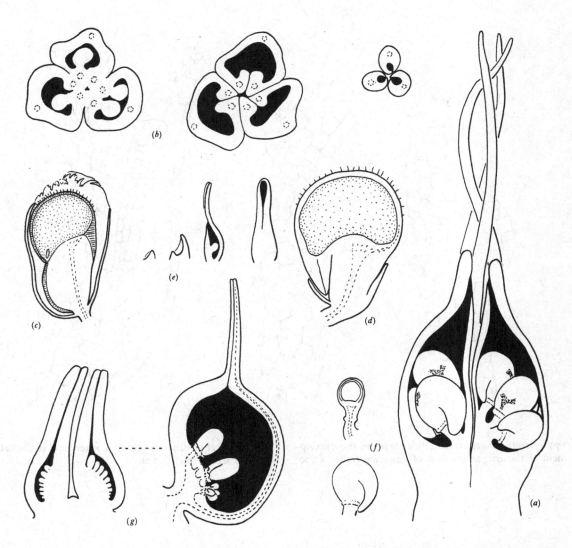

178 *Acrotrema costatum.* (*a*) ovary in l.s. shortly after pollination, the chalazal patch of v.b. hatched, ×15; (*b*) ovary in t.s. at the base, across the ovules, and across the styles, ×15; (*c*) ripe seed with aril, in l.s., ×7; (*d*) immature seed in l.s., ×15; (*e*) developing carpels, mag.; (*f*) ovules with incipient aril at the time of pollination, one in transmedian l.s., ×15; (*g*) *Tetracera indica*, immature ovary in l.s. with the ovules in basipetal sequence, ×15; young follicle in l.s., ×5.

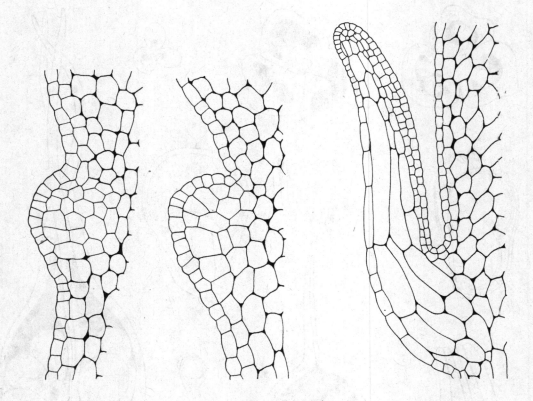

179   *Acrotrema costatum*. Three stages in the development of the aril; at the time of flowering (left), × 400; shortly after fertilisation (centre), × 400; and in the half-grown seed (right), × 200.

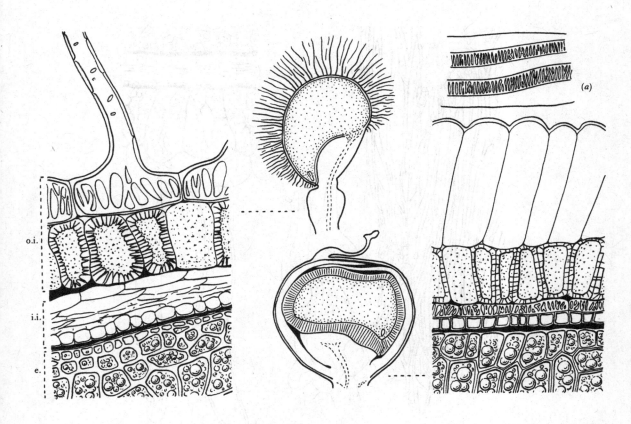

o.i.

i.i.

e.

(a)

180 *Dillenia indica* (left); mature seed in l.s., with mucilage-hairs, exarillate, ×5; seed-coats in t.s., with the base of a mucilage-hair, ×225. *Davilla sp.* (Manaus; right); mature seed with aril in the follicle, ×8; seed-coats in t.s., with well-developed tegmen, ×225; (a) rows of crystals in the outer epidermal wall of the aril, ×500.

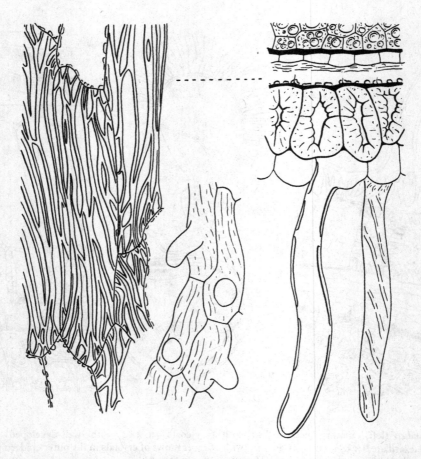

181 *Dillenia philippinensis*. Part of the seed-coat in t.s., with sclerotic endotesta, tracheidal exotegmen, and endo-sperm; the tracheidal exotegmen in surface-view; the immature exotesta with incipient hairs, × 225.

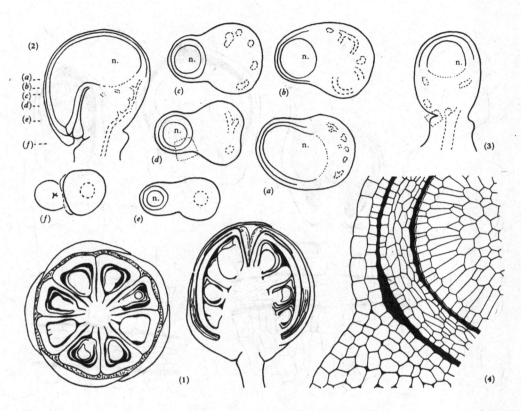

182 *Dillenia suffruticosa*. (1) Ripe fruit just before dehiscence, in t.s. and l.s., × 1½. (2) Ovule in l.s. and t.s. at various levels *a–f*, × 30. (3) Ovule in transmedian l.s., × 30. (4) Part of (2*d*), × 225.

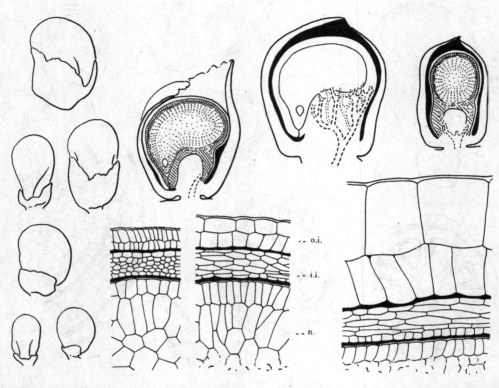

183 *Dillenia suffruticosa*. Developing seeds and mature seed (in median and transmedian l.s.), ×7. Diagram of the vascular supply of the seed. Wall of the ovule and of young seeds, ×225.

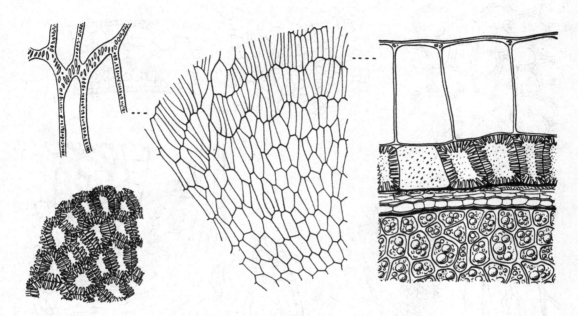

184 *Dillenia suffruticosa*. Seed-coats in t.s., with oily endosperm, × 225. Exotestal facets, × 45. Crystals in the outer parts of the radial walls of the exotesta, × 400. Endotestal facets, × 100.

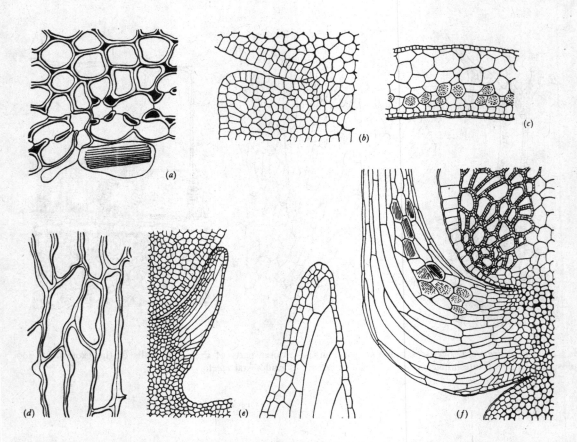

185 *Dillenia suffruticosa*. (*a*) Endocarp in t.s., with unicellular raphid hair, ×225; (*b*) aril-rudiment in l.s. at the time of flowering, ×225; (*c*) mature aril in t.s., with raphid-cells, ×50; (*d*) epidermis of the aril in surface-view, ×400; (*e*) aril of a young seed in l.s. (×50) and the margin of an older aril in l.s., ×225; (*f*) attachment of the aril to the funicle of the ripe seed, ×50.

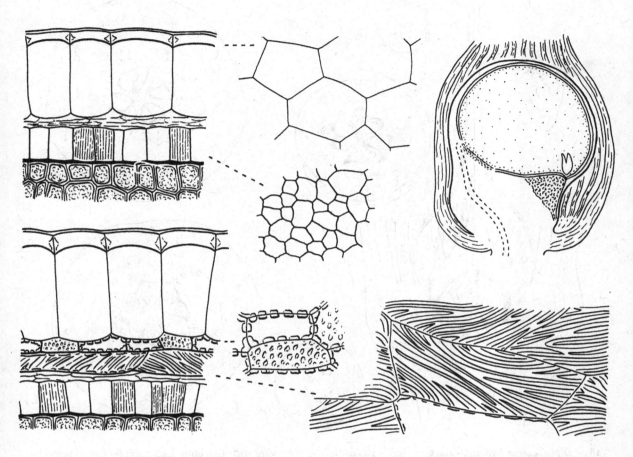

186 *Tetracera indica*. Seed in l.s. with the basal part of the aril, the broad hypostase and the sclerotic part of the subhilum near the micropyle stippled, × 12. Seed-coats in t.s.; (above) on the convex side of the seed without sclerotic endotesta or tracheidal exotegmen; (below) near the chalaza, with typical Dilleniaceous construction, × 225.

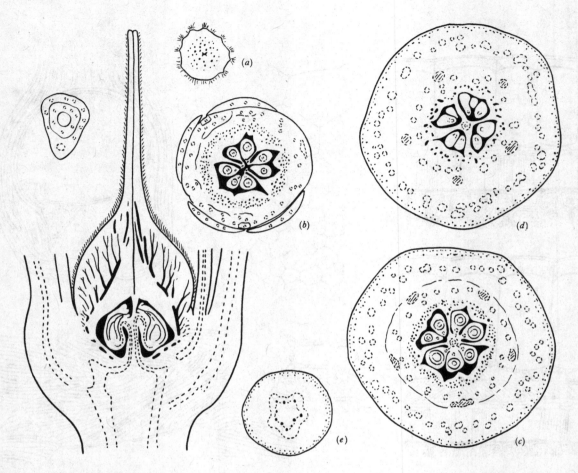

187 *Dipterocarpus hasseltii*. Ovary in l.s., ×10; ovule in t.s., ×25. (*a*) Style in t.s., ×25; (*b*) ovary in t.s. near the base, ×10; (*c*) ovary in t.s. across the middle, ×25; (*d*) ovary with 4 loculi in t.s., ×25; (*e*) flower-pedicel in t.s., ×10. Resin-canals and loculi in black.

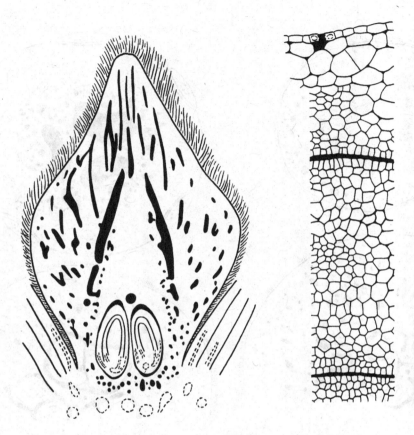

188 *Dipterocarpus obtusifolius*. Ovary in tangential l.s., × 10. Ovule-wall in t.s., with procambial strand in o.i. and i.i., and stoma of o.e. (o.i.), × 225. Resin-canals and loculi in black.

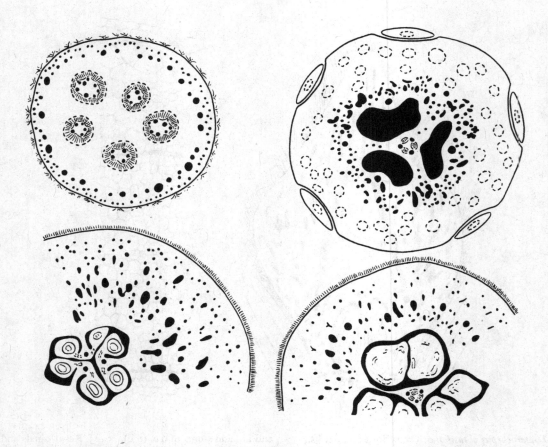

189 *Dipterocarpus obtusifolius*. Flower-pedicel in t.s.; ovary in t.s. at the base (with petal and stamen v.b.), in the middle (trilocular), and in the upper part (with parietal placentation), × 25. Resin-canals and loculi in black.

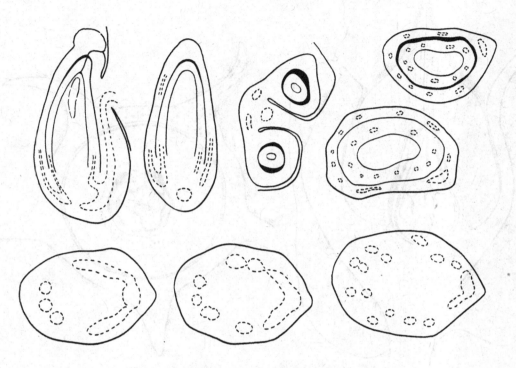

190 *Dipterocarpus obtusifolius.* Ovule in l.s. and t.s. at
various levels to show the origin of the tegmic v.b., × 25.

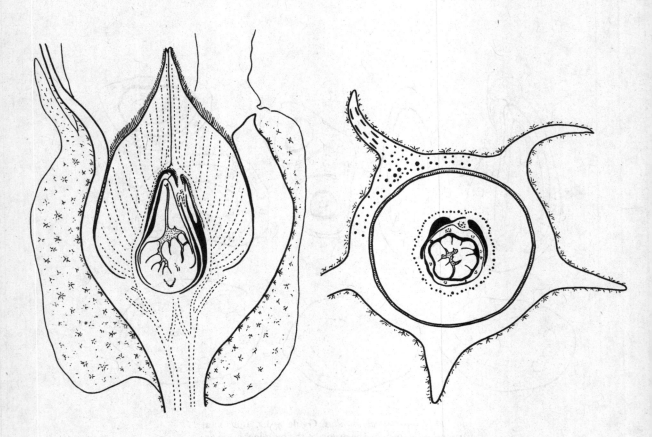

191 *Dipterocarpus zeylanicus.* Half-grown fruit in l.s. (with indications of v.b.) and in t.s. (with indications of resin-canals accompanied by v.b.), × 5.

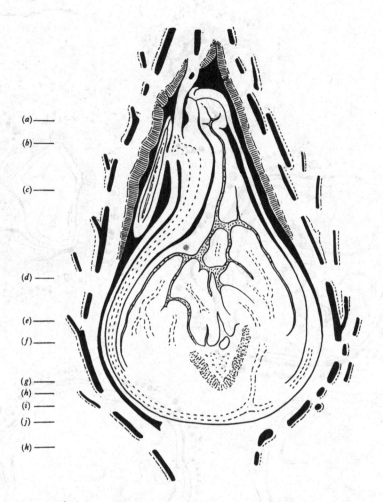

(a)
(b)
(c)
(d)
(e)
(f)
(g)
(h)
(i)
(j)
(k)

192 *Dipterocarpus zeylanicus.* Young seed in l.s. in the endocarp, with proliferated outgrowths of the chalaza, ×10. Resin-canals accompanied by v.b. and loculi in black. See Figs. 193 and 194 for sections at the levels (*a*)–(*k*).

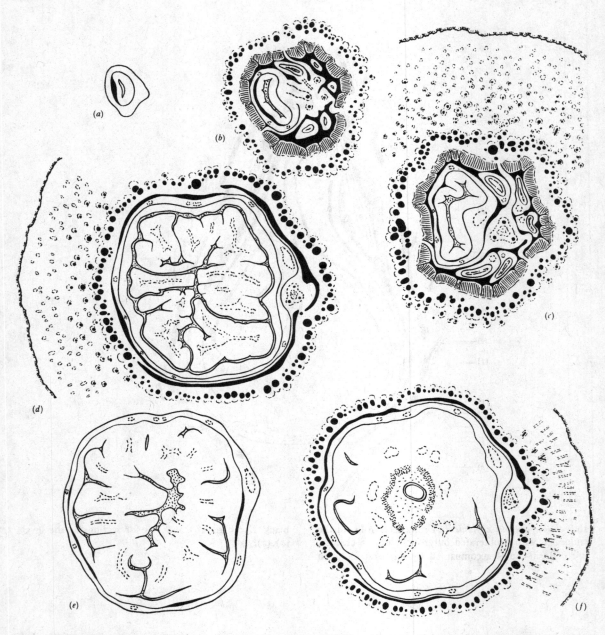

193 *Dipterocarpus zeylanicus.* The young seed of Fig. 192 in t.s. at the levels (*a*)–(*f*) ((*c*), (*d*) and (*f*) also with part of the pericarp); (*b*) and (*c*) with the endocarp lignifying into a palisade, × 10. Resin-canals accompanied by v.b. and loculi in black.

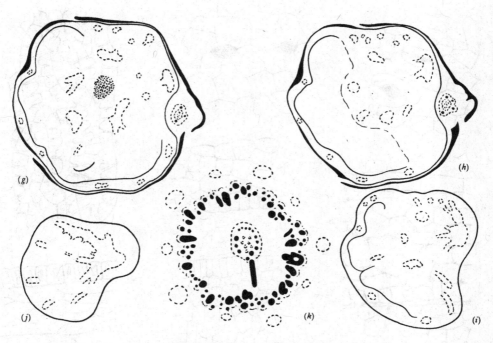

194 *Dipterocarpus zeylanicus.* As Figs. 192, 193, with
sections at the levels (g)–(k), × 10.

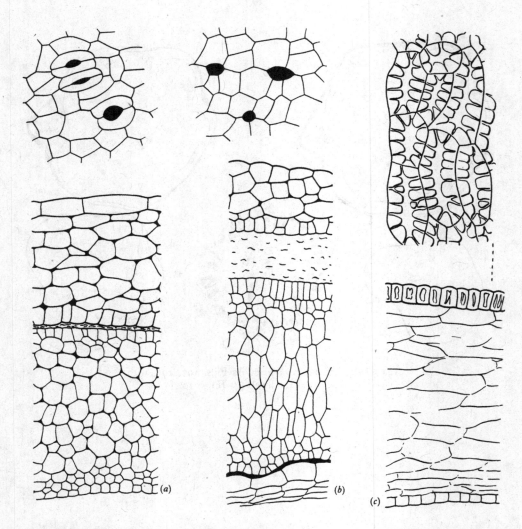

195 *Dipterocarpus zeylanicus*; testa and tegmen in t.s. from the middle of the immature seed (*a*) and from near the micropyle (*b*); stomata of the exotesta, × 225. *Dipterocarpus sp.* (Malaya); (*c*) tegmen of the mature dried seed, with short palisade-cells, × 225; palisade-cells in surface-view, × 500.

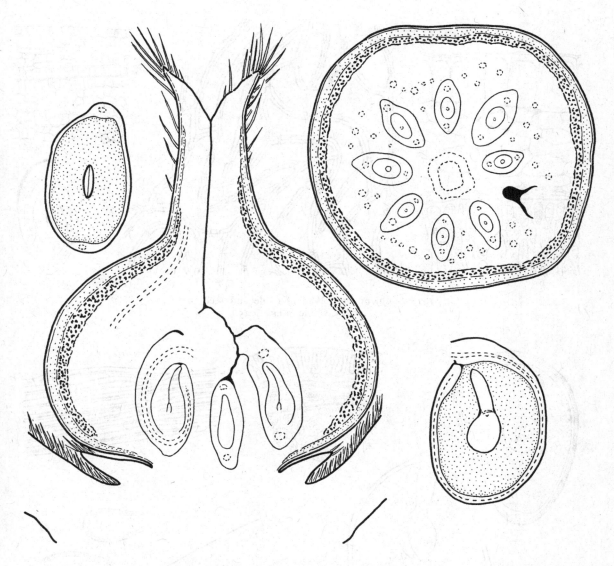

196 *Diospyros lotos*. Ovary in l.s. (slightly oblique), showing the stylar canal to the ovules, the stone-cells (heavily stippled), and the thick-walled cells in the outer part of the ovary-wall, ×25. Ovary in t.s. near the base, ×25. Ripe seeds in l.s. and t.s., ×8.

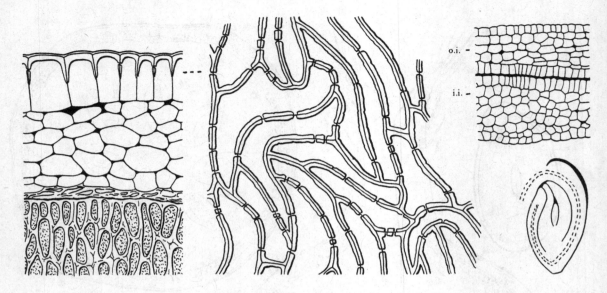

o.i. -

i.i. -

197   *Diospyros lotos.* Wall of ovule and seed in t.s.,
×225. Ovule in l.s., ×25.

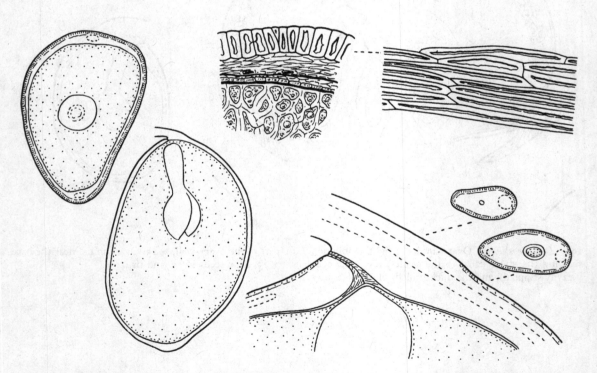

198   *Diospyros mollis.* Seed in l.s., ×5; in t.s. across the
radicle, ×10. Micropylar end of the seed with woody
endostome in median l.s., ×25, and sections across the
endostome, ×10. Seed-coat in t.s. with endosperm and a
trace of tegmen, with exotestal facets, ×225.

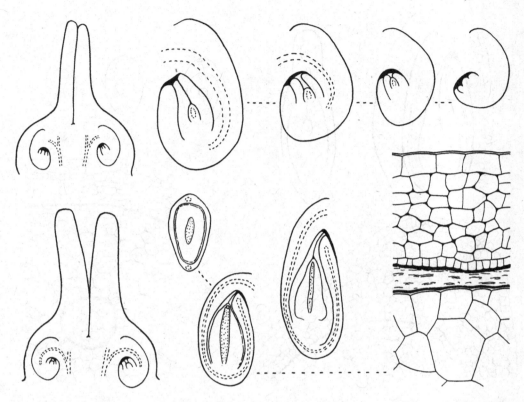

199 *Diospyros oblonga.* Developing ovaries in l.s., × 30. Developing ovules in l.s., to show the chalazal thickening, × 50. Young seeds to show the growth of the nucellus, in l.s. and t.s. (left), × 7; in l.s. (right), × 15. Seed-coat of immature seed with the tegmen degenerating, and large nucellar cells, × 225.

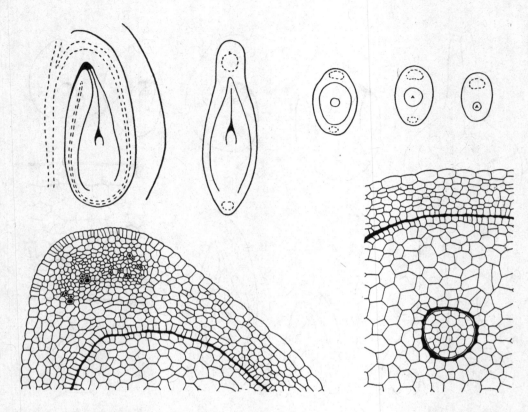

200 *Diospyros oblonga.* Ovule in median and trans-median l.s. and in t.s. at different levels, × 30. Raphe and side of ovule in t.s., to show the massive construction, × 225.

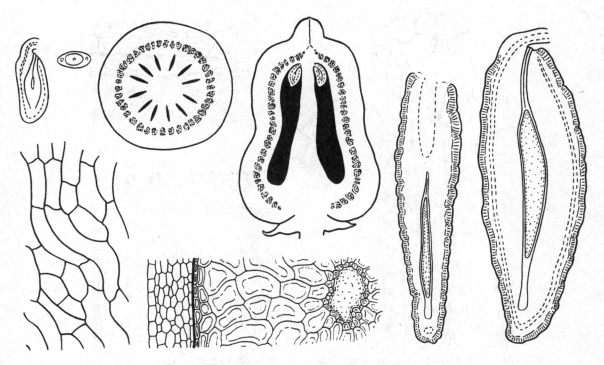

201 *Diospyros quaesita*. Immature fruit in l.s. and t.s., showing the basipetally extended loculi and the cylinder of sclerotic masses near the endocarp, × ⅔. Ovule soon after fertilisation in l.s. and t.s., × 10. Immature seeds in median and transmedian l.s., × 10. Seed-coat of the ovule soon after fertilisation with enlarged mucilaginous cells of the tegmen, × 225. Epidermal cells of the testa of immature seeds, × 115.

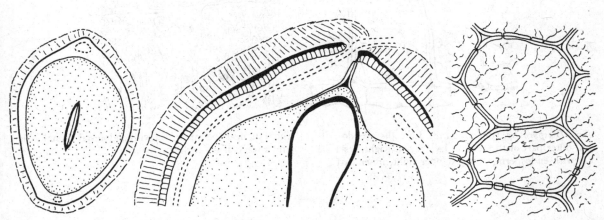

202. *Diospyros sp.* (Philcox 4178). Mature seed in t.s., invested by the pseudo-aril (endocarp), × 5. Micropyle in median l.s., showing the pseudo-aril, the exotestal palisade, and the trace of the tegmen at the endostome, × 10. Exotestal facets, × 225.

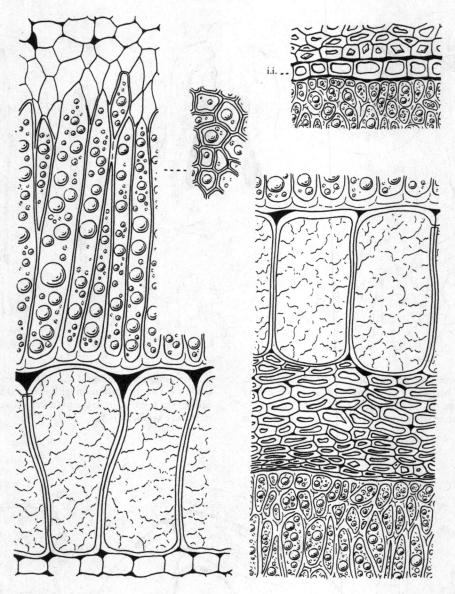

203 *Diospyros sp.* (Philcox 4178). Left, the pseudo-aril and outer part of the testa in l.s., ×225. Right, testa in t.s. with endosperm, ×225. Upper right, inner part of the testa near the micropyle with endotestal crystal-cells and a trace of the tegmen, ×225.

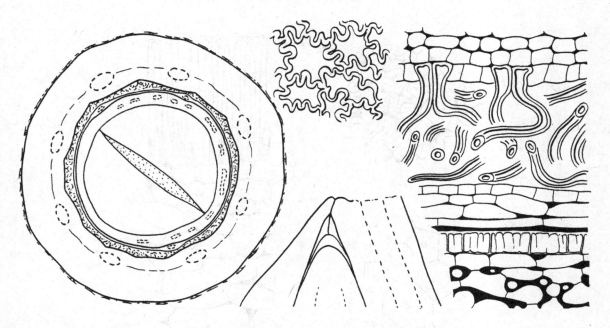

204 *Elaeagnus latifolius*. Perianth-tube (with 8 v.b.), the pericarp (3 v.b.) and the seed with thick cotyledons in t.s., ×8. Inner part of the perianth-tube with hairs, pericarp (3–4 cells thick), and the outer part of the testa in t.s., ×225. Exotestal facets, ×500. Micropyle in l.s. with a trace of the endostome and nucellus, ×25.

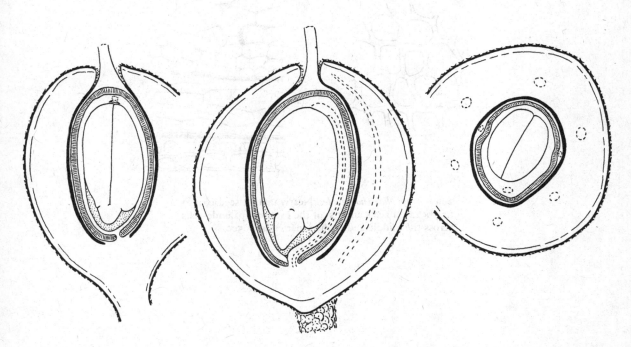

205 *Hippophae rhamnoides*. Fruits in median and trans-median l.s. and in t.s.; perianth-tube with narrow small-celled cortex and wide large-celled inner part with v.b.; pericarp thin, with 2 v.b.; testa with palisade (striate), ×10.

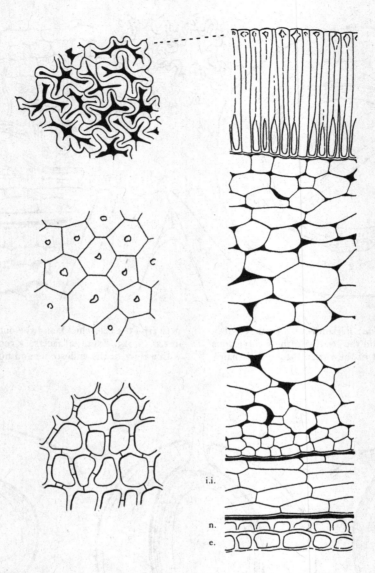

206 *Hippophae rhamnoides.* Nearly mature seed–coat in
t.s., × 225. Facets and t.s. of the exotestal palisade–cells
across the middle and at the inner ends, × 500.

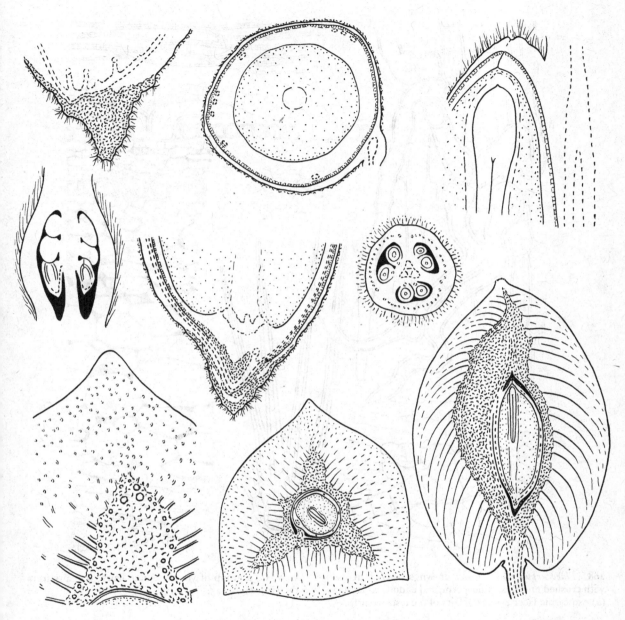

207 *Elaeocarpus edulis.* Ovary in l.s. and t.s., ×25. Fruit (? with immature embryo) in l.s. and t.s., ×2; angle of the fruit, showing v.b. in the woody endocarp, ×5. Seed in t.s., l.s. and tangential l.s. of the chalaza, and the micropyle in l.s., ×10.

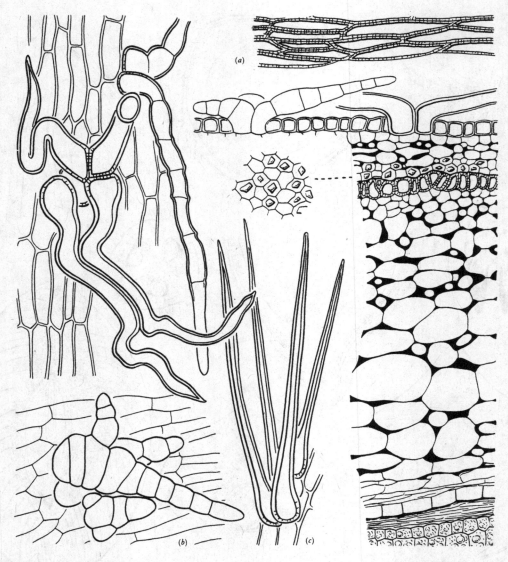

208 *Elaeocarpus edulis.* Fully grown seed-coat in t.s. with crushed nucellus, endosperm, and endotestal facets; (*a*) exotegmic facets, × 225. Hairs of the testa in surface-view; (*b*) a group of young septate hairs; (*c*) hairs from the chalaza, × 225.

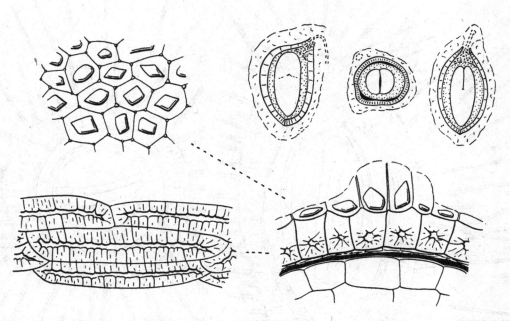

209 *Muntingia calabura*. Seeds in t.s., median and transmedian l.s., embedded in the mucilage of the exotesta, × 50. Mature seed-coat in t.s. with endotestal crystal-cells, exotegmic fibres, and outer layer of the endosperm, × 500. Surface-view of the endotesta and exotegmen, × 500.

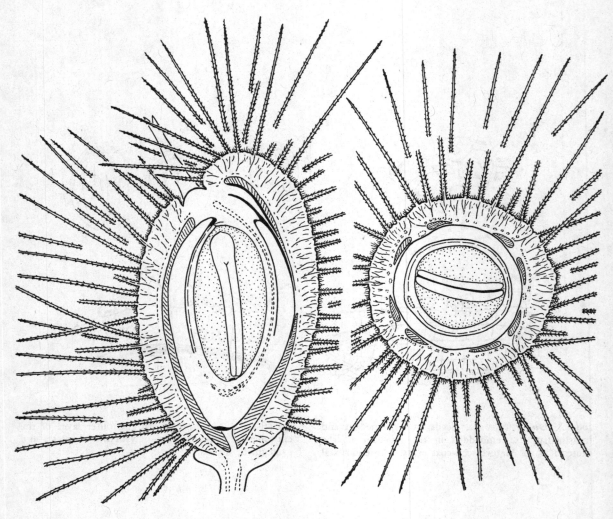

210   *Sloanea alnifolia*. Fruit in l.s. and t.s.; endocarp
with the valve-tissue hatched, × 5.

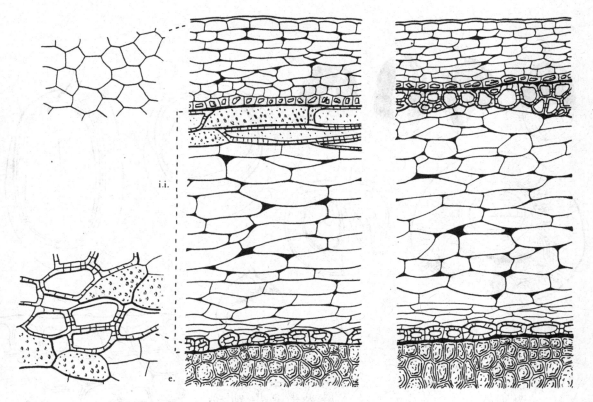

i.i.

e.

211 *Sloanea alnifolia*. Seed-coats in t.s. (right) and l.s. (centre) with endosperm; endotesta as crystal-cells, exotegmen fibrous, endotegmen sclerotic, × 225.

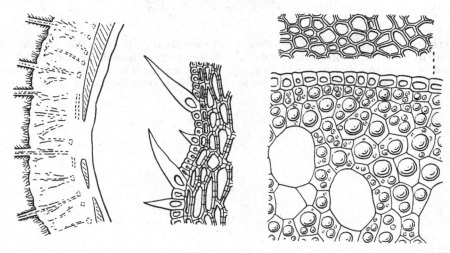

212 *Sloanea alnifolia*. Pericarp in t.s. (left) with the valve-tissue hatched, and with v.b. to the bristles, × 12. Surface of a bristle (centre), × 225. Aril (right) in t.s., × 225.

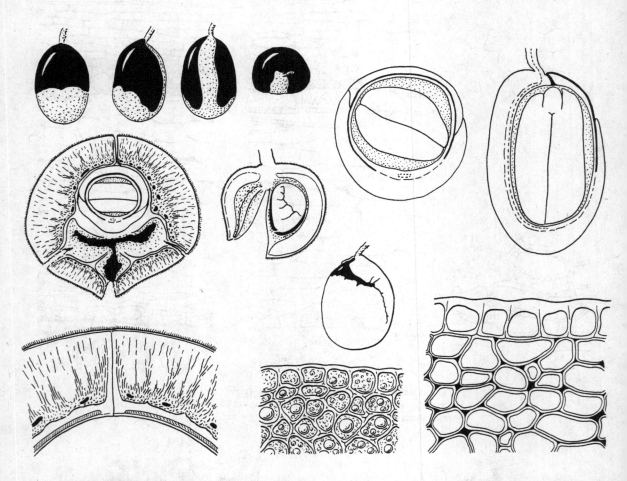

213 *Sloanea javanica*. Seed in four aspects, with the aril removed to show its attachment along the raphe and over the chalazal end, × 1. Ripe fruit dehiscing, in l.s., × ½; in t.s., × 1. Part of the pericarp at a loculicidal fissure to show the endocarp valve (hatched) and woody mesocarp with hollows (from disintegrating v.b.), × 3. Seed with aril, × 1. Seeds in l.s. and t.s., × 2. Surface of the aril (with oily cells) and testa in t.s., × 400.

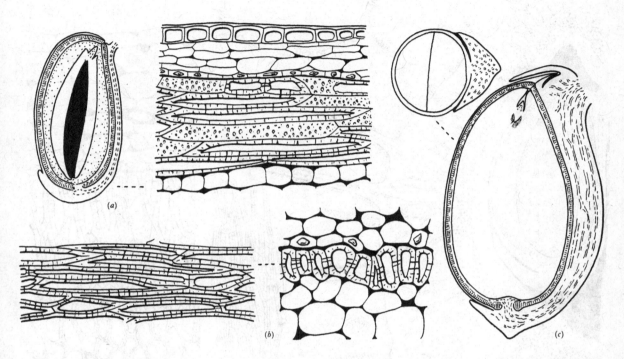

214 (a) *Sloanea celebica*: seed in l.s., ×3; testa and outer part of tegmen in l.s., ×225. (b) *S. javanica*: testa in t.s. and exotegmen in surface-view, ×225. (c) *Gonystylis forbesii*: seed in l.s., ×3 and t.s., ×2.

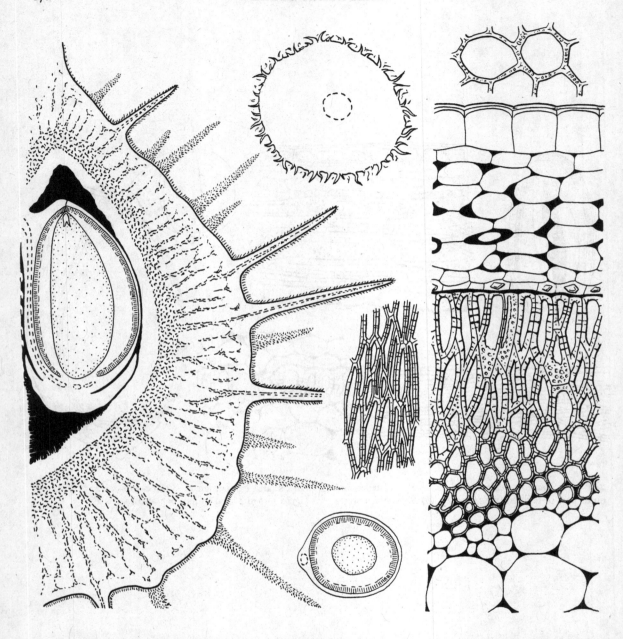

215 *Sloanea sigun*. Part of the immature, but fully grown, capsule in l.s., showing the woody endocarp with fibro-vascular bundles traversing the mesocarp and entering the spines, the loculus hairy in the lower part and with abortive ovules next the raphe of the seed, × 5. Spine in t.s., × 60. Immature seed in t.s., with the woody part of the tegmen hatched, × 5. Testa and outer part of the tegmen in t.s., with exotestal facets, × 225. Exotegmen in surface-view at lower magnification to show the cell-proportions, × 120.

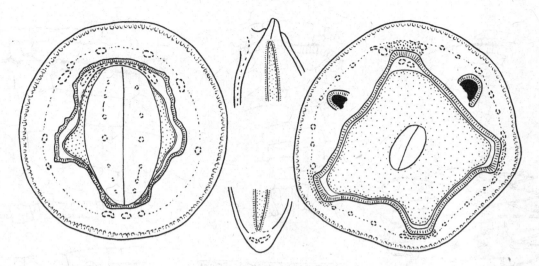

216 *Erythroxylon*. Ripe drupes of *E. moonii* (left) with large embryo and thin endosperm, and of *E. coca* (right) with two undeveloped sterile loculi, small embryo, and thick endosperm, in t.s., × 10; hypodermal layer of oil-cells, ring of v.b. separating the two layers of the meso-carp, thin woody endocarp (hatched), and strips of sclerotic tissue at the angles of the endocarp in *E. coca*. Median l.s. of the micropyle of an immature seed of *E. coca*, with transmedian l.s. of the chalaza, showing the prominent endostome, endothelial endotegmen, and double v.b. at the chalaza, × 25.

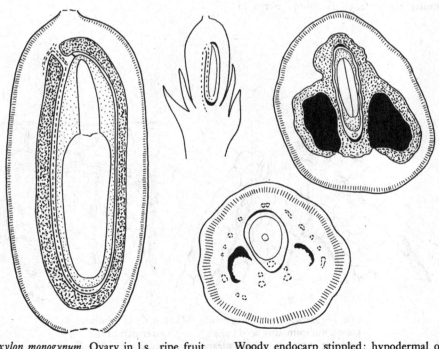

217 *Erythroxylon monogynum*. Ovary in l.s., ripe fruit in l.s. and t.s., × 10; young fruit in t.s. (below), × 25. Woody endocarp stippled; hypodermal oil-cells of the pericarp striated.

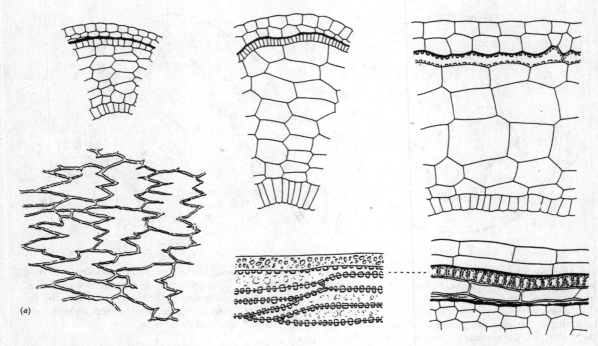

218 *Erythroxylon monogynum.* Wall of ovule and of developing seed in t.s., and of mature seed in t.s. and l.s. with fibrous exotegmen, × 225; exotegmic fibres in surface-view, × 500; (*a*) facets of the outer hypodermal cells of the tegmen, × 225.

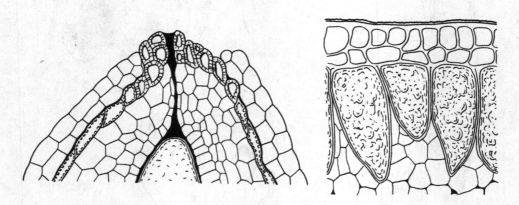

219 *Erythroxylon monogynum.* Micropyle of fully grown but immature seed in l.s., × 225. Outer part of the pericarp in t.s., with large oleaginous cells, × 225.

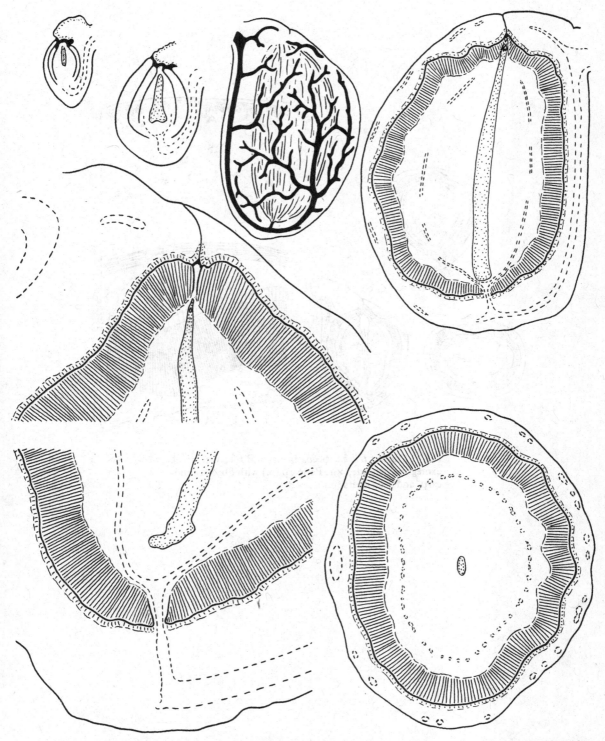

220 *Aleurites moluccana*. Ovule and very young seed in l.s., with obturator, × 25. Mature seed to show the vascular supply on one side of the testa and the tegmic bundles from the chalaza, × 1. Fully grown but immature seed in l.s., the embryo as yet microscopic, the thin-walled endotestal palisade and the woody exotegmic palisade striated, × 2. Similar seed in t.s., × 3. Micropyle and chalaza of the same seed in l.s., × 8.

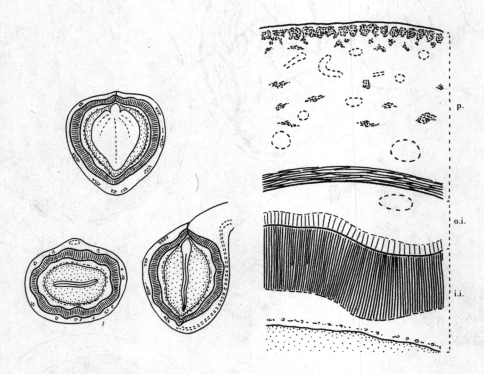

221 *Aleurites triloba*. Seeds in l.s. and t.s., with thick exotegmic palisade, × 1. Pericarp (p.) with fibrous endo-carp, and seed-coats in t.s., × 8.

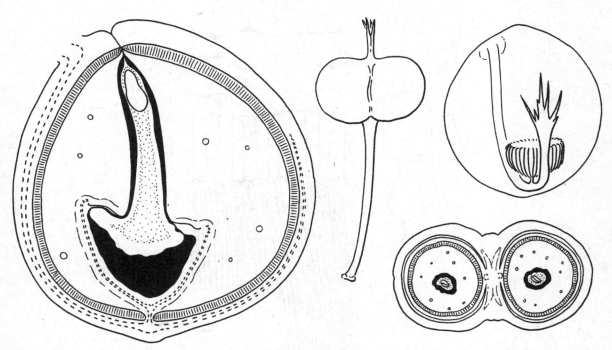

222   *Cleidion javanicum*. Seed (immature) in l.s., with a hollow round the contracted nucellus, showing the pachychalaza, ×8. Fruit, ×1; section, ×2. Vascular supply to the seed, mag.

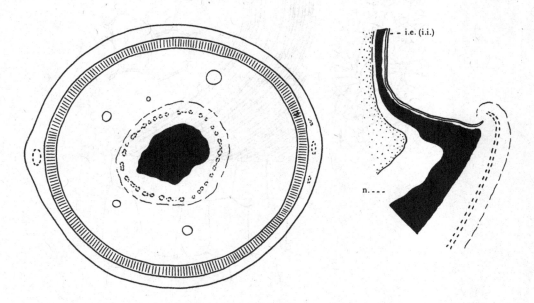

223   *Cleidion javanicum*. Seed in t.s. across the pachychalaza, ×8. Pachychalazal rim at the junction with the tegmen, ×25.

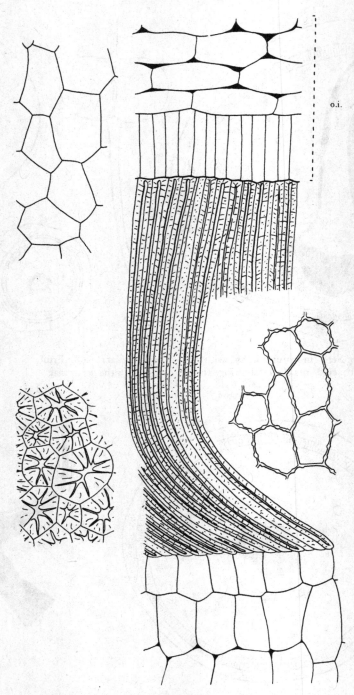

o.i.

224 *Cleidion javanicum*. Tegmen in l.s. in the middle part of the seed, × 225; exotestal facets, × 225; facets of exotegmen and endotegmen, × 500.

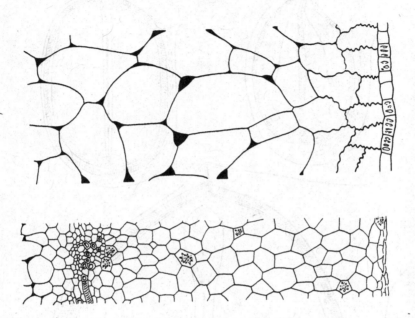

225 *Cleidion javanicum*. Above, inner part of the tegmen in l.s. with the endotegmen on the right, × 225.

Below, pachychalazal wall in t.s., with the collapsing inner surface on the right, × 225.

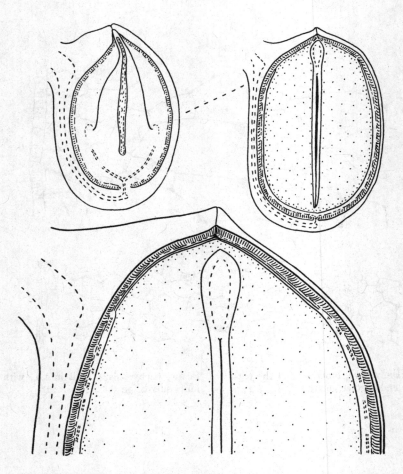

226 *Cleidion javanicum* (material from Java). Immature seed with tegmic pachychalaza in l.s. (× 12) and mature seed filled with endosperm but with a trace of the tegmic pachychalaza except in the region of the radicle (× 3). Micropylar end of the seed to show the junction of the tegmen and nucellus with the pachychalaza, × 8.

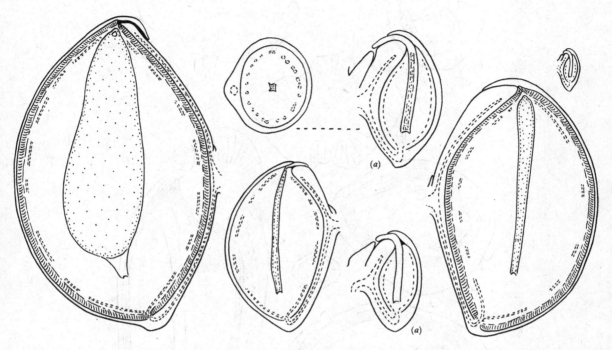

227 *Croton laevifolium*. Development of the ovule with tegmic v.b. into the seed; nucellus narrow, exsert through the micropyle, soon substituted by endosperm; seed reaching full-size with thick tegmen and fully formed exotegmic palisade, but the endosperm partly developed and the embryo microscopic, × 12. (*a*) × 25.

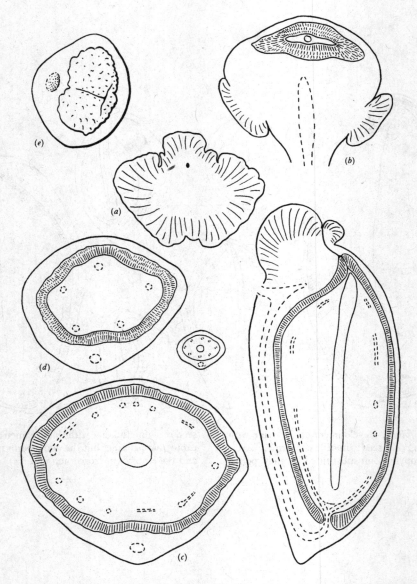

228 *Croton sp.* (Ceylon). Seed (fully grown but immature) in l.s. and t.s. at various levels: (*a*) at the arillostome; (*b*) below the arillostome; (*c*) across the middle; (*d*) just above the chalaza, × 18. Apex of seed (*e*), × 10. Ovule (centre) in t.s., × 18.

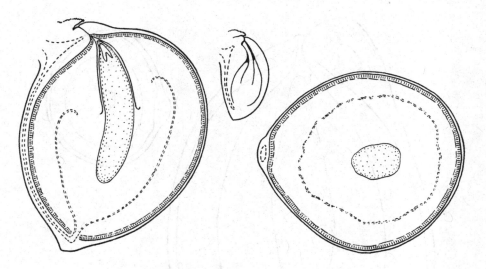

229 *Dimorphocalyx glabellus*. Ovule in l.s., × 25. Immature seeds in l.s. and t.s. with the exotegmen striated, × 12.

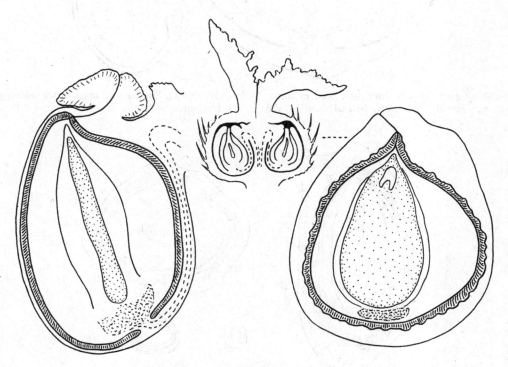

230 *Euphorbia pilosa* (left). Fully grown but immature seed in l.s., × 25. *Mercurialis perennis* (right). *Ovary*, × 25, and fully grown but immature seed, × 12, in l.s.

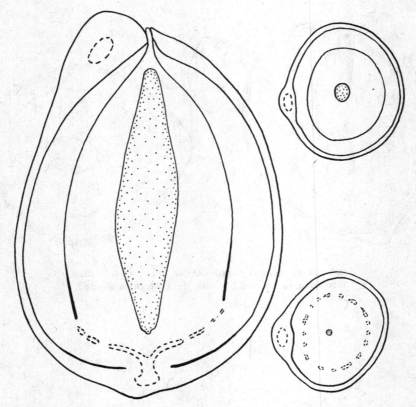

231 *Gelonium glomerulatum*. Immature seed in trans-median l.s., showing the pachychalazal base of the nucellus, × 25. Similar seed in t.s. at the micropylar and chalazal ends, × 12.

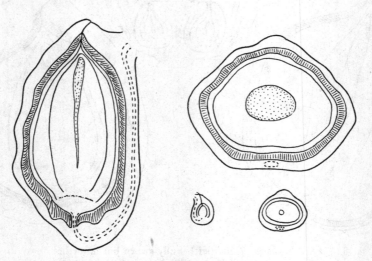

232 *Homalanthus populneus*. Ovule, young seed and fully grown but immature seed in l.s. and t.s., × 10.

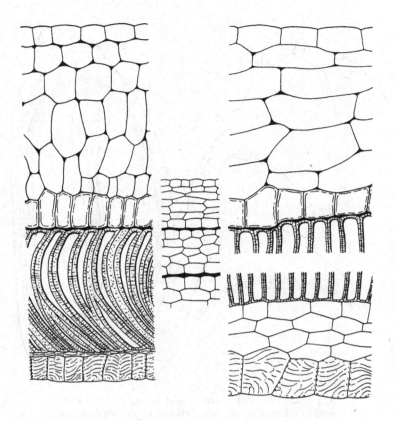

233 *Homalanthus populneus*. Wall of ovule in t.s., of immature seed in l.s. (right) and mature seed (left) in l.s., × 225.

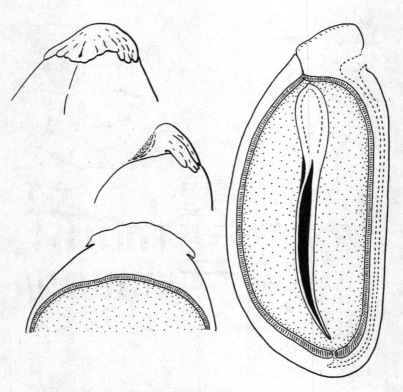

234 *Jatropha curcas.* Mature seed in l.s., × 5. Apex of
seed in micropylar and side-view and in transmedian l.s.
to show the slight aril, × 5.

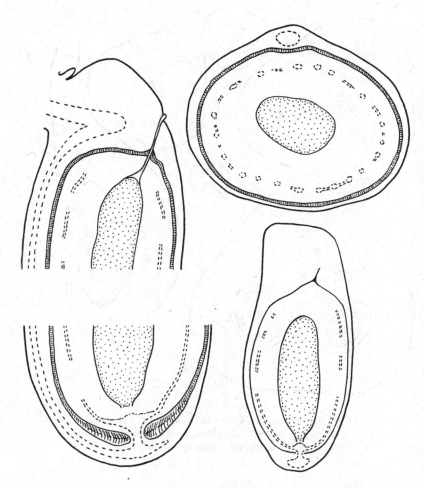

235 *Jatropha curcas*. Immature seed in l.s. and t.s. and in transmedian l.s., to show the tegmic v.b. and the nucellar remains at the micropyle, × 8.

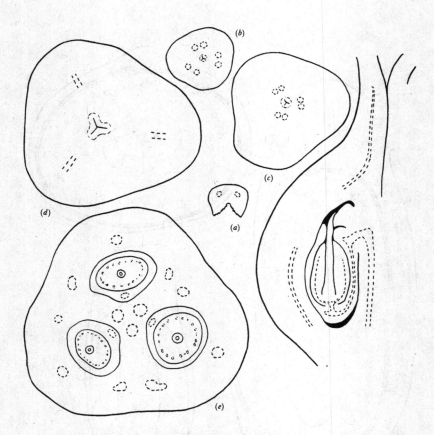

236 *Jatropha curcas.* Ovary in l.s. and in t.s. at various levels: (*a*) across a stigma; (*b*) across the style; (*c*), (*d*) at the apex of the ovary; (*e*) across the ovules; ×25.

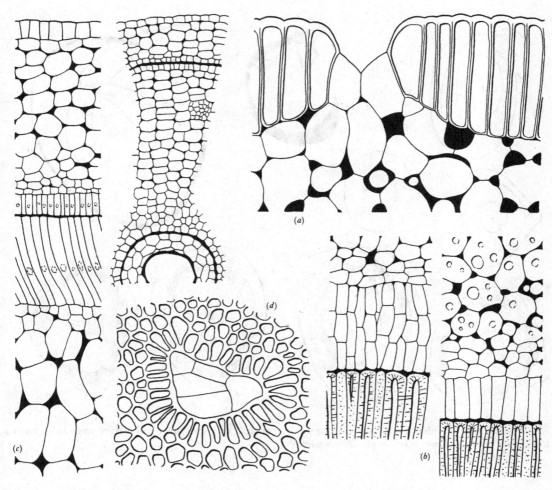

237 *Jatropha curcas.* Ovule-wall and seed-coats in t.s.: (a) mature seed with outer part of the testa (with a colourless thin-walled patch in the exotesta) and (b) inner part of the testa at the junction with the exotegmic pali-sade (with multiple endotesta in the chalazal region); (c) immature seed; (d) surface-view of a thin-walled colourless patch on the exotesta; × 225.

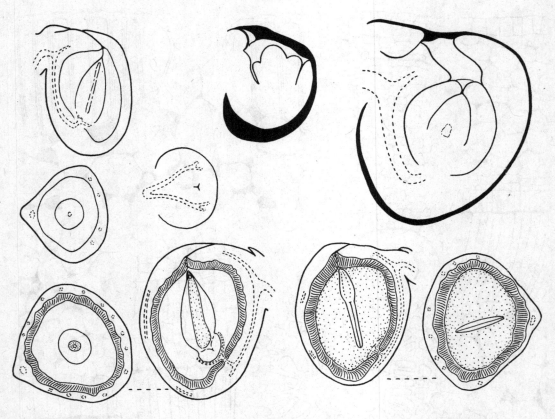

238 *Mallotus barbatus*. Ovules at anthesis in l.s., t.s., and micropylar view to show v.b., ×30. Developing ovules, ×75. Immature and mature seeds in l.s. and t.s., ×7.

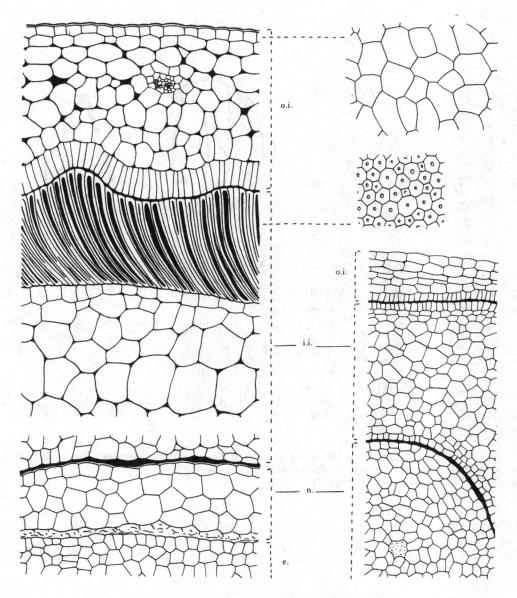

239 *Mallotus barbatus.* Walls of ovule (right), ×225 and
mature seed (left), ×75.

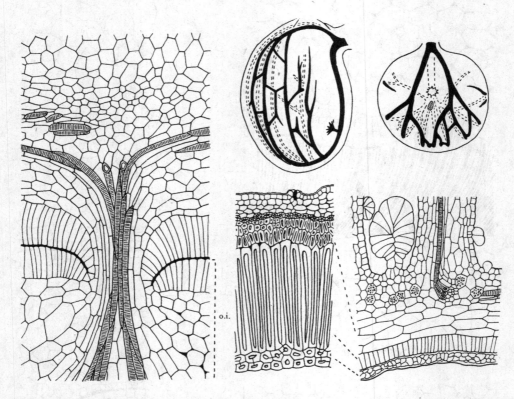

240 *Mallotus barbatus*. Chalaza of immature seed in l.s., ×225. Vascular supply of the seed in side-view and micropylar view (micropyle hatched), mag. Ovary-wall in t.s., ×225; mature pericarp with sclerotic tissue in l.s., ×75.

241 *Mallotus sp.* (Ceylon). Fruit in t.s. at the level of the micropyles, ×10. Ovule in l.s., ×25. Fully grown but immature seed in l.s. and t.s., ×10.

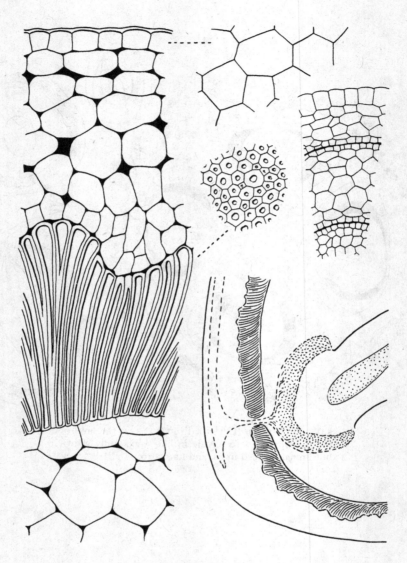

242 *Mallotus sp.* (Ceylon). Wall of ovule and seed-coats
in t.s., × 225. Chalaza in l.s., × 25.

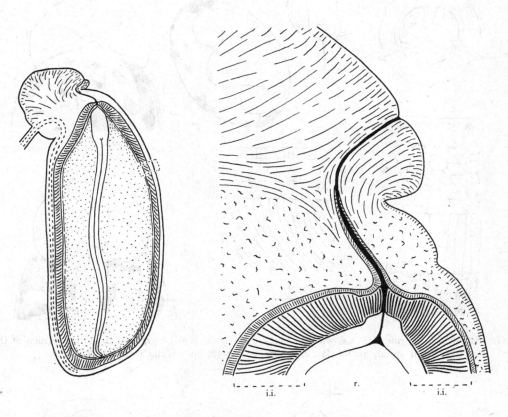

243　*Manihot utilissima*. Seed in l.s., × 6. Micropyle and arillostome in l.s., × 50. r. radicle.

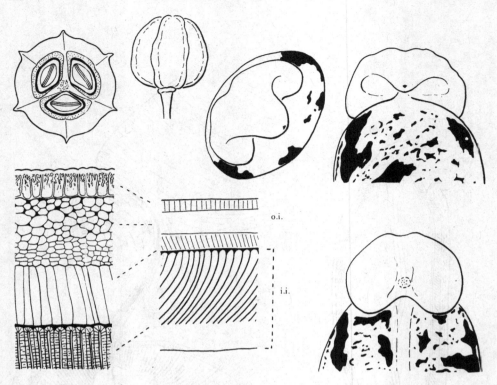

244 *Manihot utilissima*. Fruit, ×1; t.s. with the cocci stippled, ×1½. Aril from apical, micropylar, and hilar views, ×6. Diagram of testa and tegmen of the seed, ×30; microscopic details, ×90.

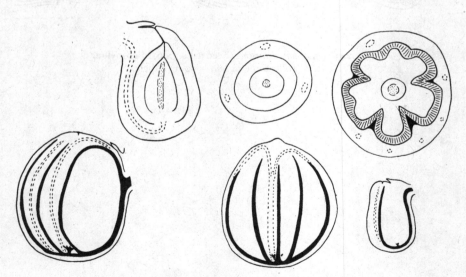

245 *Melanolepis multiglandulosa*. Ovule in l.s. and t.s., ×30. Fully grown but immature seed in t.s., ×7. Vascular supply to the seed and ovule, mag.

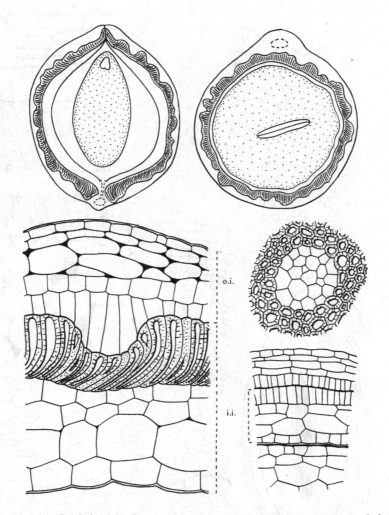

o.i.

i.i.

246 *Micrococca mercurialis.* Seed in l.s. (immature embryo) and in t.s., × 25. Seed-coats in l.s. of mature and young seeds, with a tangential section across a 'pit' of thin-walled endotesta surrounded by the exotegmic palisade, × 225.

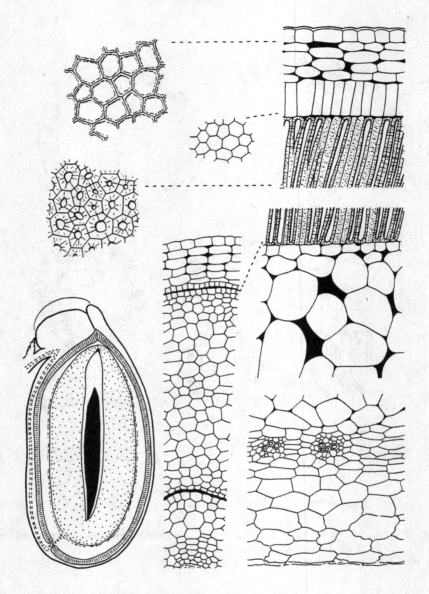

247 *Ricinus communis*. Seed, nearly mature, in l.s., ×8.
Wall of ovule in t.s. and of the seed-coats in the pachychalazal region with internal v.b. (lower right), ×225.

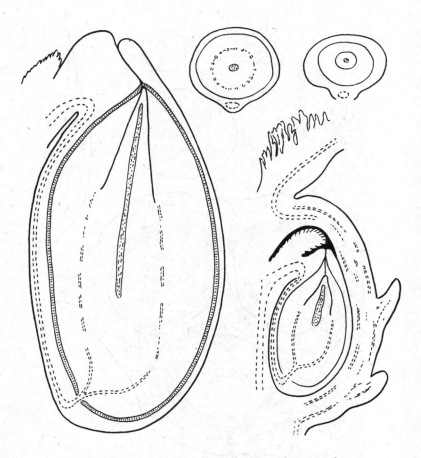

248 *Ricinus communis.* Ovule in l.s. and t.s. (pachy-chalazal region and nucellar region) and young seed in l.s. (with conical nucellus above the pachychalaza), × 25.

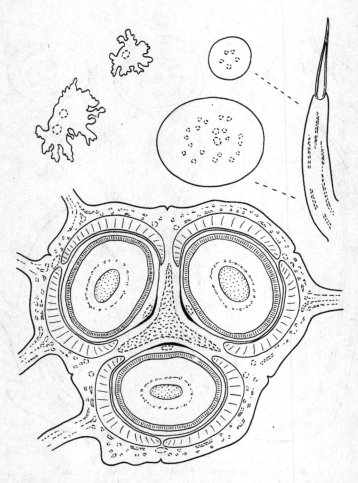

249 *Ricinus communis*. Fully grown but immature fruit in t.s., × 8. Lobulate stigmatic arm and base in t.s., × 25.

Spine of ovary and sections of the enlarged spine of the fruit to show v.b. and hair-tip, × 25.

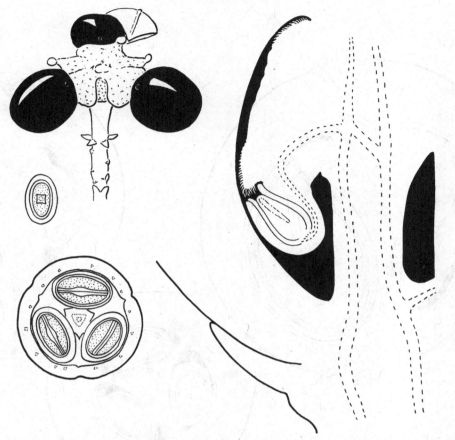

250 *Aporosa frutescens.* Ripe fruit in t.s., and ripe seed in t.s. across the 4-angled hypocotyl, × 2. Dehisced fruit, the elevated sarcotestal seeds throwing off the cocci, with 3 abortive ovules, × 2. Ovule in l.s. soon after fertilisation, × 30.

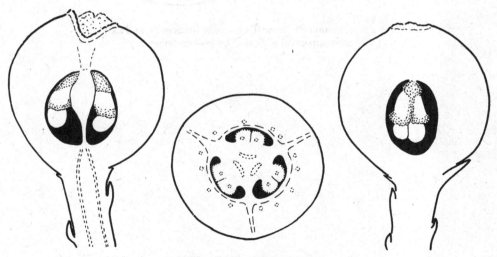

251 *Aporosa frutescens.* Female flowers in l.s. and t.s. at anthesis; tangential l.s. (right) to show the paired ovules with massive obturators; × 7.

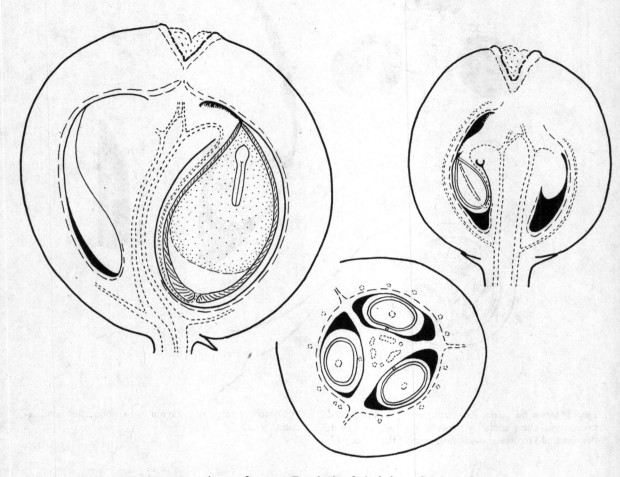

252   *Aporosa frutescens*. Developing fruits in l.s. and t.s.;
endocarp-cocci indicated by broken lines; × 7.

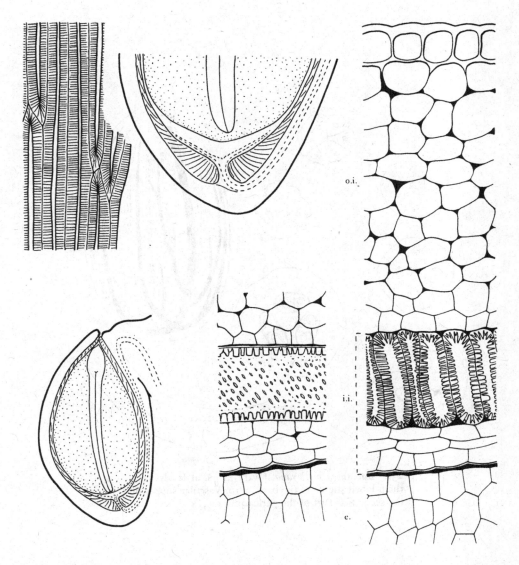

253   *Aporosa frutescens*. Mature seed in l.s., with sarcotesta, × 6. Chalaza in median l.s., showing the direction of the exotegmic fibres, × 15. Seed-coat in t.s., the
tegmen in l.s., exotegmic fibres in surface-view, × 225.

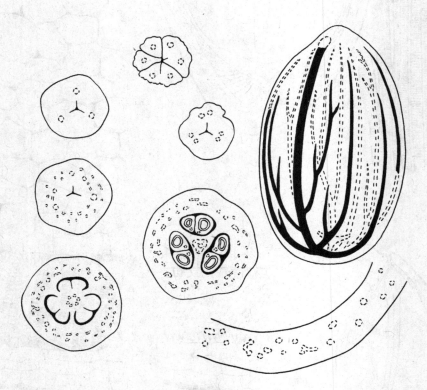

254 *Baccaurea motleyana*. Ovary in t.s. at levels from
the 6-lobed stigma to the base, × 10. Vascular supply of
the seed, × 4. Part of the pericarp in t.s., × 5.

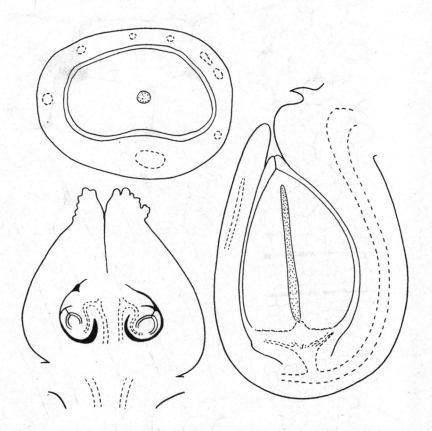

255 *Baccaurea motleyana*. Ovary in l.s. and developing
seeds in l.s. and t.s., × 18.

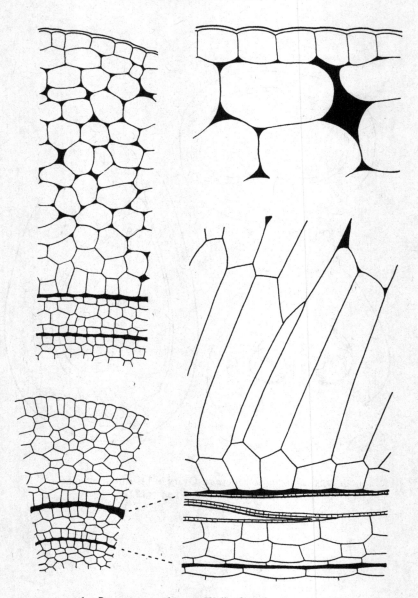

256 *Baccaurea motleyana.* Wall of ovule, young seed, and almost mature seed, with sarcotesta and fibrous exotegmen, × 225.

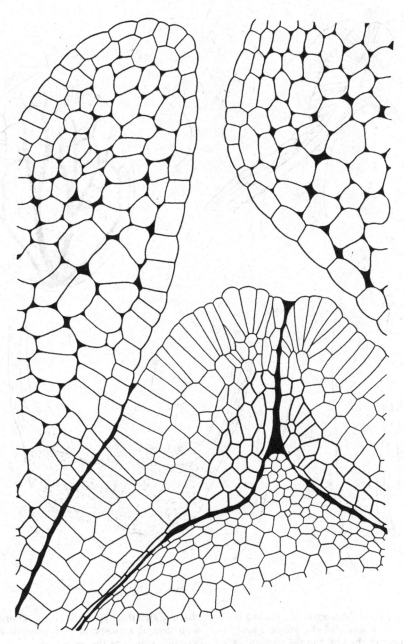

257   *Baccaurea motleyana.* Micropyle of the young
seed.   ×225.

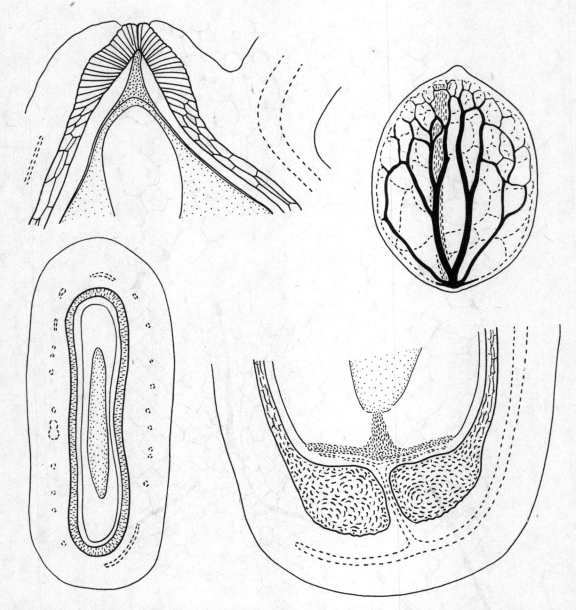

258 *Baccaurea sp.* (S. 29458). Micropyle and chalaza in median l.s. of a fully grown seed and of a younger seed, showing the arrangement of the fibres in the tegmen, of the mass of sclerotic cells adjoining the tegmen round the chalaza, and the hypostase, × 17. Abortive seed in t.s. with central endosperm and massive nucellus, × 10. Vascular supply of the testa, × 5.

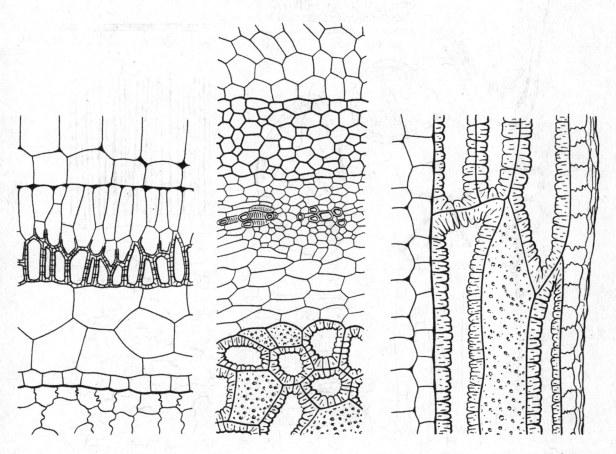

259 *Baccaurea sp.* (S. 29458). Left, tegmen of an immature seed in t.s., showing the thickening of the fibres from the inner tissue outwards, the nucellar cells collapsing, × 225. Centre, t.s. of the hypostase-tissue, the vascular supply of the chalaza, and the mass of sclerotic cells in the chalaza, × 225. Right, tegmen in l.s. with 3 layers of fibres, the inner tissue collapsed, × 225.

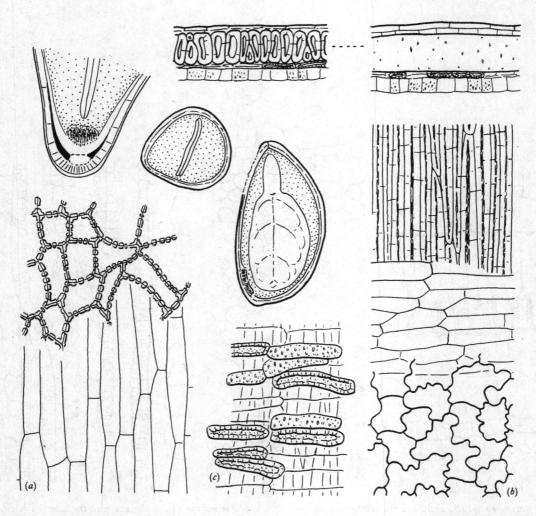

260 *Bischofia javanica*. Seeds in l.s. and t.s., ×10. Chalaza in transmedian l.s., with the fibrous exotegmen separated from the tegmen, ×25. Seed-coats in t.s. and l.s.: (*a*) inner pitted cells of the tegmen overlying the nucellar remains, ×500; (*b*) exotestal cells (undulate) and endotestal, overlying the fibrous exotegmen, ×225; (*c*) sclerotic cells of the mesotegmen with outlines of the exotegmic fibres, ×225.

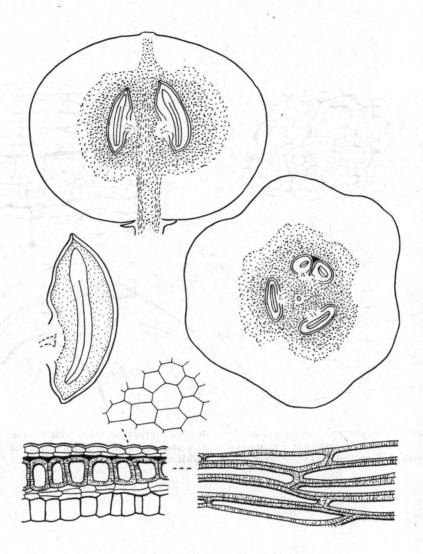

261   *Cicca acida*. Fruit in l.s. and t.s., with woody endocarp, ×5. Seed in l.s., ×10. Seed-coats in t.s., ×225.

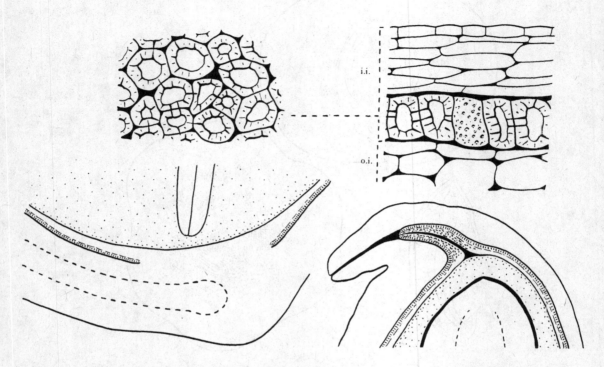

262 *Drypetes laevis*. Micropylar and chalazal regions of a mature seed, ×25; exotegmen hatched; endostome with sclerotic mesophyll. Seed-coats in t.s., showing the tegmen (uppermost layer) with 6 layers of thin-walled cells and the cuboid lignified exotegmic cells, with two cell-layers of the inner testa, ×225.

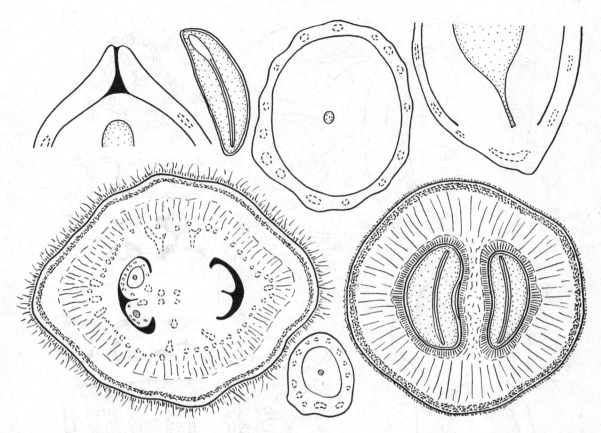

263 *Drypetes macrostigma.* Ovary in t.s. of the upper part with an outer zone of sclerotic tissue and radiating fibro-vascular procambial strands in the mesocarp external to the longitudinal v.b., × 12. Fruit in t.s. with one seed in each loculus; pericarp with sclerotic zone in the exocarp, radiating fibro-vascular strands in the mesocarp, and woody endocarp (hatched), × 2. Mature seed in l.s., × 4. Ovule (lower centre) in t.s., × 25. Half-grown seed in t.s., and l.s. of micropylar and chalazal regions (upper row), × 12.

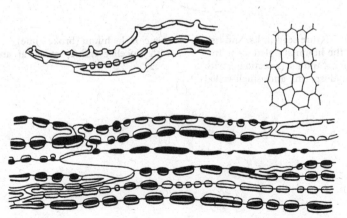

264 *Galearia filiformis.* Surface-view of the exotegmic fibres and of the hexagonal cells of the endotegmen, × 225; some fibres thin-walled; the air-spaces between the short arms of the fibres shown in black.

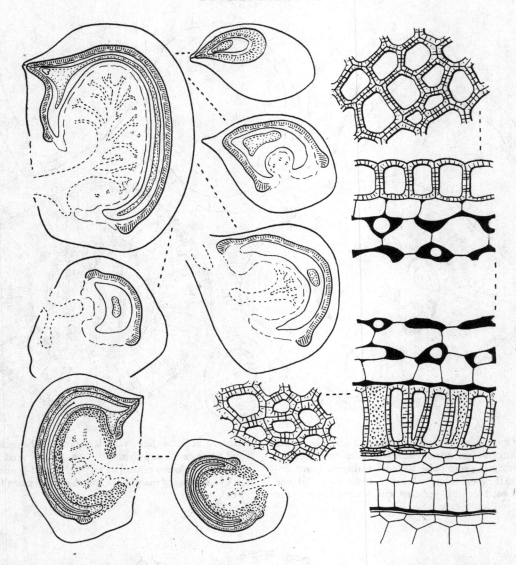

265 *Glochidion zeylanicum*. Mature seed in l.s. and t.s., with the sclerotic tissue of the hilum stippled, × 7. Immature but full-sized seed in l.s. and t.s. (4 sections), with large nucellus, incipient endosperm, and small-celled tissue of the hilum (broken line), × 12. Seed-coats in t.s. with nucellar tissue, exotestal and exotegmic facets, × 225.

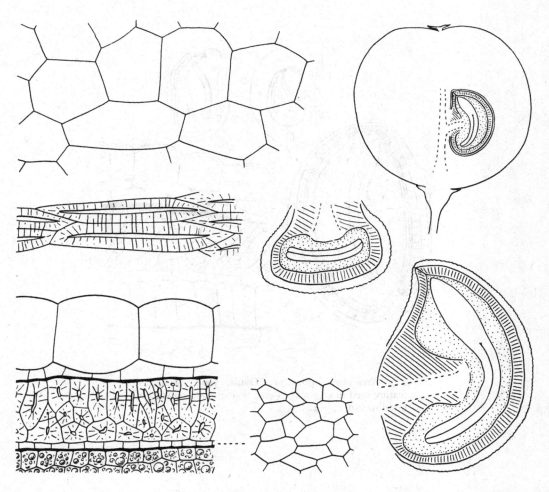

266 *Securinega leucopyrum*. Fruit in l.s., ×8. Seeds in l.s. and t.s., the lignified part of the seed-coat hatched, ×25. Seed-coats in t.s., with exotestal facets and exotegmic fibres in surface-view, ×225.

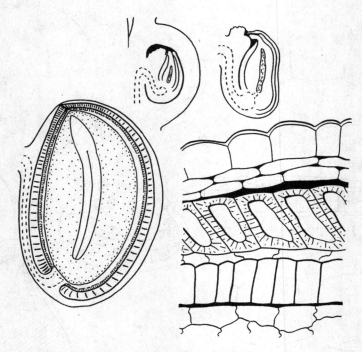

267 *Securinega suffruticosa*. Ovule, young seed and mature seed in l.s., ×25. Testa, tegmen, and remains of nucellus in t.s., ×225.

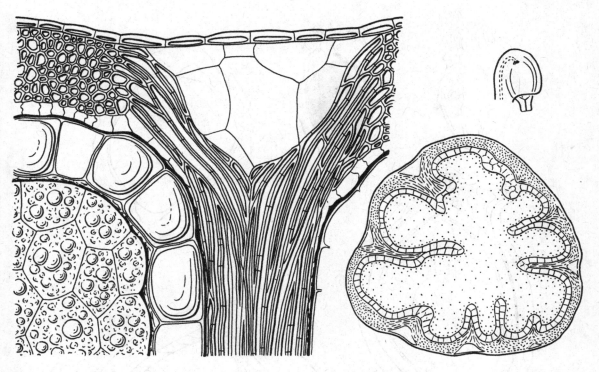

268 *Eupomatia laurina.* Ovule with short o.i., × 30. Seed in t.s., with testal fibres and tegmic oil-cells, × 18.

Rumination at the outer end with adjacent fibrous testa, tegmen with large oil-cells, and endosperm, × 225.

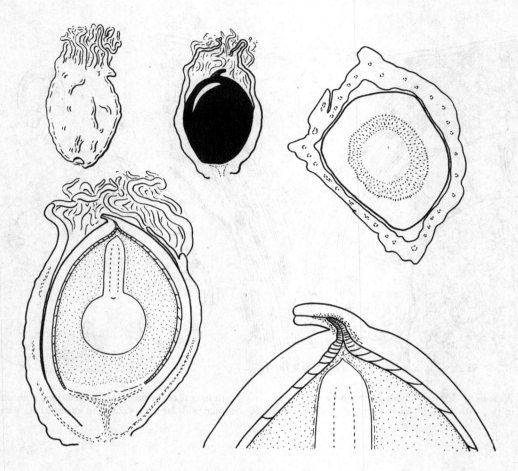

269 *Casearia rugulosa*. Seed in aril and with the aril partly removed, ×5. Seed in l.s., ×10; in t.s. at the chalaza, ×25. Micropyle in l.s., ×25.

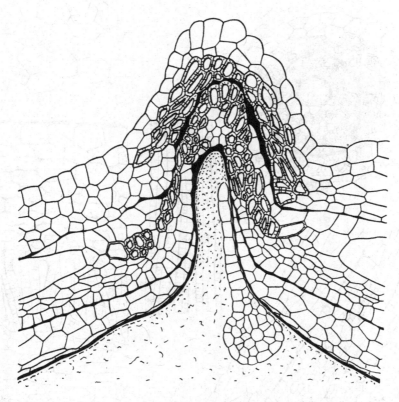

270   *Casearia rugulosa*. Micropyle in transmedian l.s. of an immature seed; exotegmic fibres thin-walled; endo-sperm protruding through the apex of the nucellus, ×225.

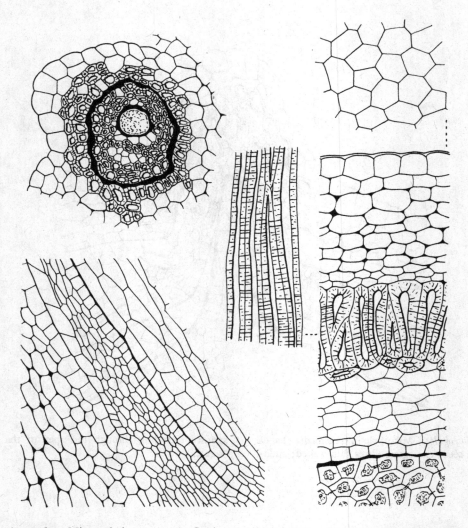

271 *Casearia rugulosa*. Micropyle in t.s., × 225. Seed-coats in t.s., with endosperm, × 225. Junction of the integuments with the nucellus and chalaza in an immature seed, × 225.

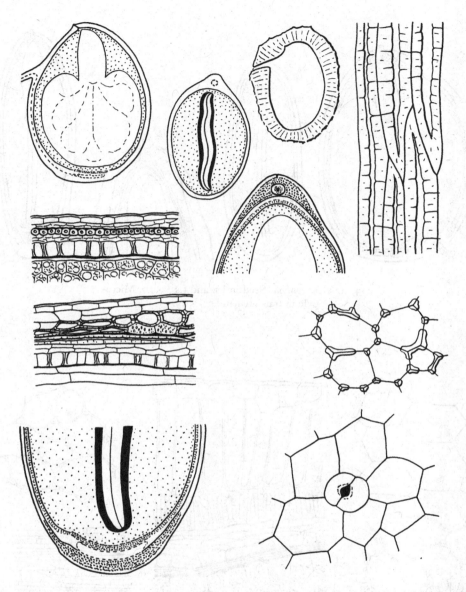

272 *Flacourtia indica.* Seeds in l.s. and t.s., × 10. Pyrene in l.s., × 5. Micropylar and chalazal ends of the seed in transmedian l.s., × 25. Seed-coats in t.s. (upper) and l.s. (lower) near the chalaza with sclerotic cells in the testa, × 225. Exotestal facets, with stoma, endotegmic facets, and exotegmic fibres, × 500.

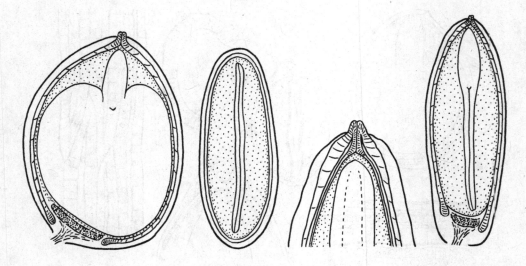

273   *Oncoba spinosa*. Seed in l.s. and t.s., × 10. Micropyle in transmedian l.s., × 25.

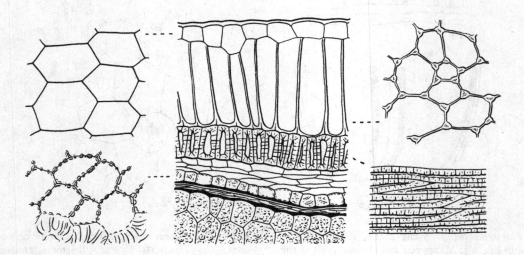

274   *Oncoba spinosa*. Mature seed-coat in t.s., with endosperm and remains of the nucellus, × 225; facets of the exotesta and fibres of the exotegmen, × 225; facets of the endotesta and endotegmen, × 500.

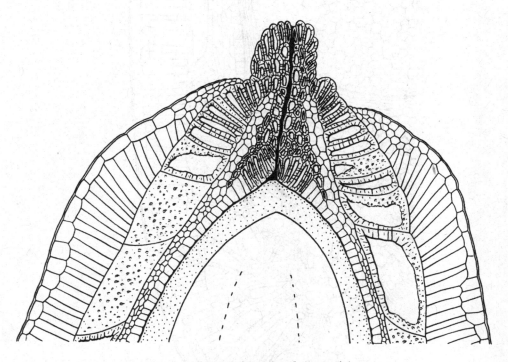

275 *Oncoba spinosa*. Micropyle of the seed in transmedian l.s., × 115.

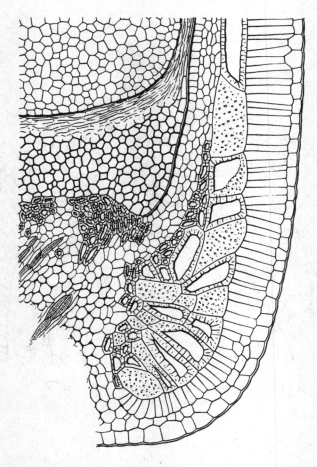

276 *Oncoba spinosa*. Chalazal region on one side of the seed in transmedian l.s., showing endosperm, nucellar remains, hypostase, sclerotic tissue in the chalaza, sclerotic thickening of the tegmen at the junction with the chalaza, and the thin-walled endotestal palisade, ×115.

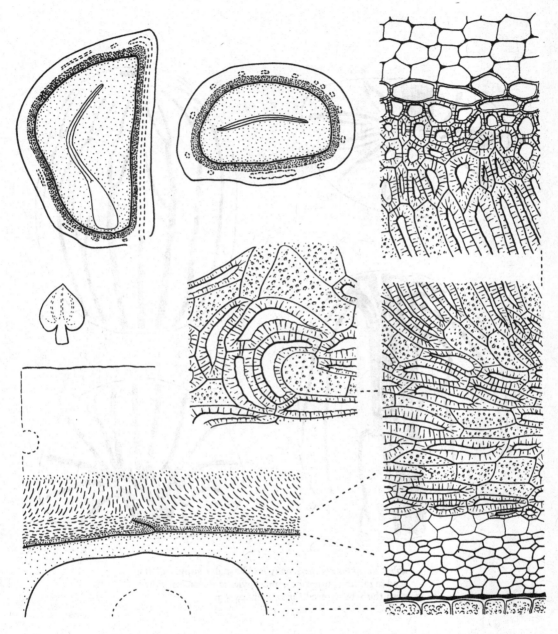

277 *Taraktogenos heterophyllus.* Seed in l.s. and t.s., with the inner sclerotic tissue of the seed-coat stippled, × 2. Embryo, × 1. Micropylar end of the seed, with the micropyle occluded in the testa, × 25; microscopic details of the inner part of the seed-coat in this region, × 225.

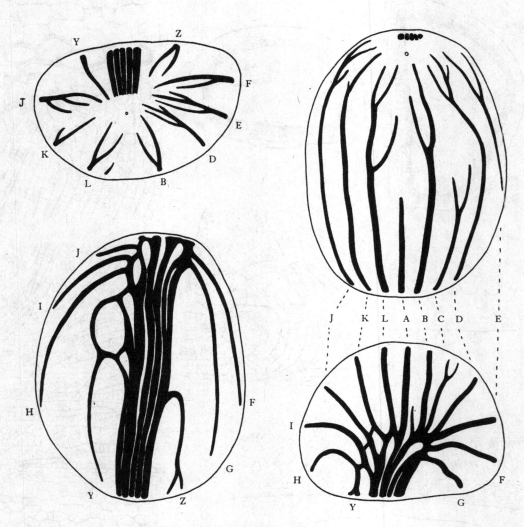

278 *Taraktogenos heterophyllus.* Vascular supply of the seed in raphe-, antiraphe-, micropylar and chalazal views, with the v.b. lettered for reference, × 3.

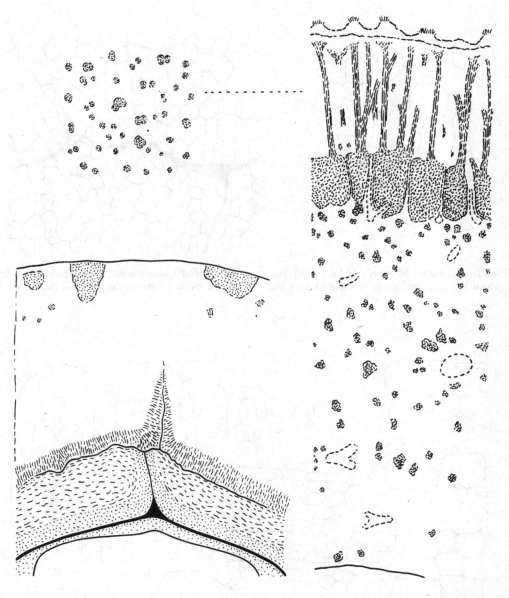

279 *Taraktogenos kurzii*. Micropylar region of the seed in l.s., showing the short limits of the free testa and tegmen, the relation between the tangentially directed sclerotic tissue of the tegmen and the radial sclerotic tissue of the inner part of the testa, and the dark patches in the outer part of the testa, × 10. Pericarp in t.s., showing v.b., sclerotic tissue (stippled and streaked), and the periderm; tangential section of the outer cortical radiating sclerotic strands, × 10.

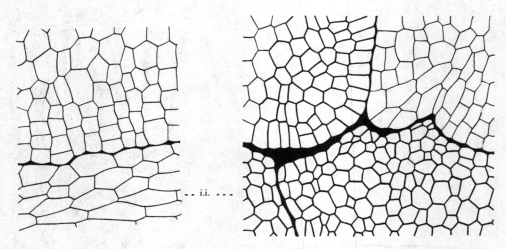

280   *Taraktogenos kurzii*. Micropyle in l.s. (right) from a fully grown but immature seed; junction of testa and tegmen (left), near the micropyle, showing the derivation of the radial sclerenchyma from the endotesta, × 225.

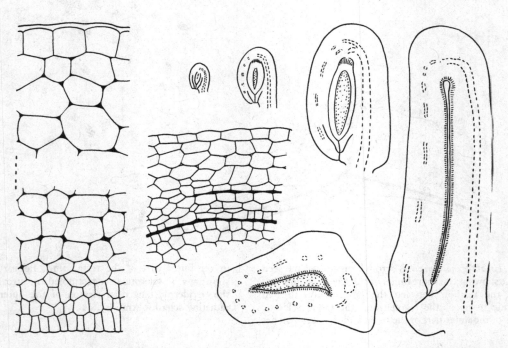

281   *Taraktogenos sp.* (Ceylon). Ovule and developing seed in l.s., the oldest in t.s., showing the pachychalazal construction with vestigial integuments, × 10. Ovule-wall in l.s. at the junction of the integuments, and the pachychalazal wall in t.s., × 225.

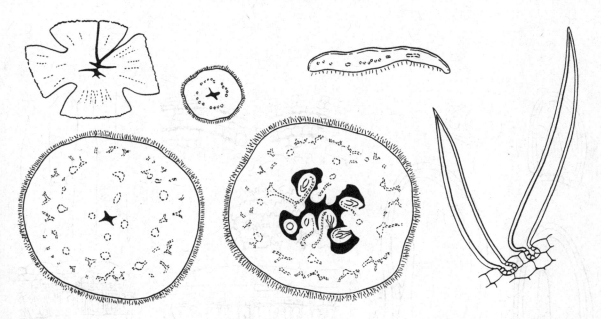

282 *Taraktogenos sp.* (Ceylon). Ovary in t.s. at several levels from the stigmata to the placentas, × 10. Stigmatic lobe in t.s., × 10. Hairs of the ovary, × 225. Ovary-wall with cortical v.b. in two rows.

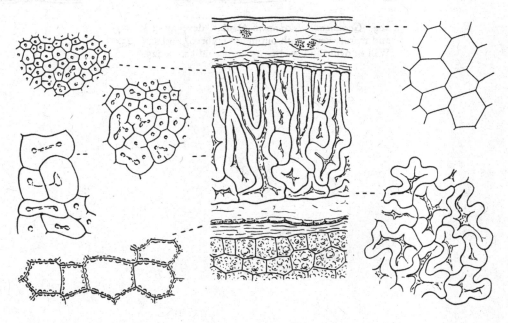

283 *Biebersteinia multifida.* Seed-coats in t.s. with endosperm, the testa thin-walled, the exotegmen sclerotic with radiating arms, the endotegmen persistent, × 225; facets of the endotegmen, × 500.

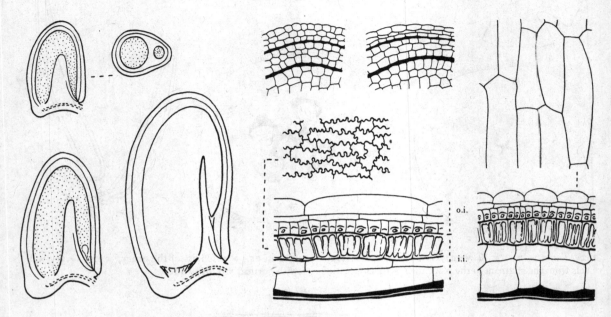

284 *Geranium robertianum.* Ovules, developing seeds, and mature seed (with outline of embryo), in l.s., ×25. Wall of ovule and seed-coats in t.s. and l.s., ×225.

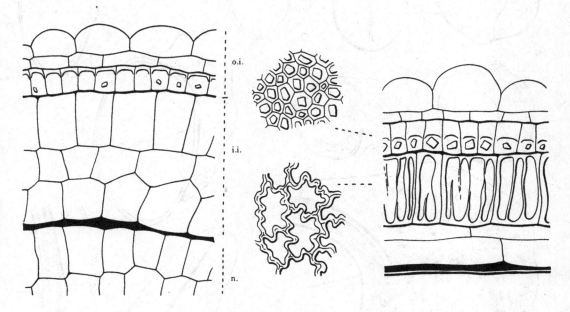

285 *Geranium robertianum*. Immature seed-coats (left)
and mature (right) in t.s.; crystal-cells of the endotesta
developing thick outer walls, × 400.

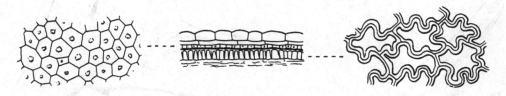

286 *Monsonia angustifolia*. Seed-coats in t.s.,
× 225; facets, × 500.

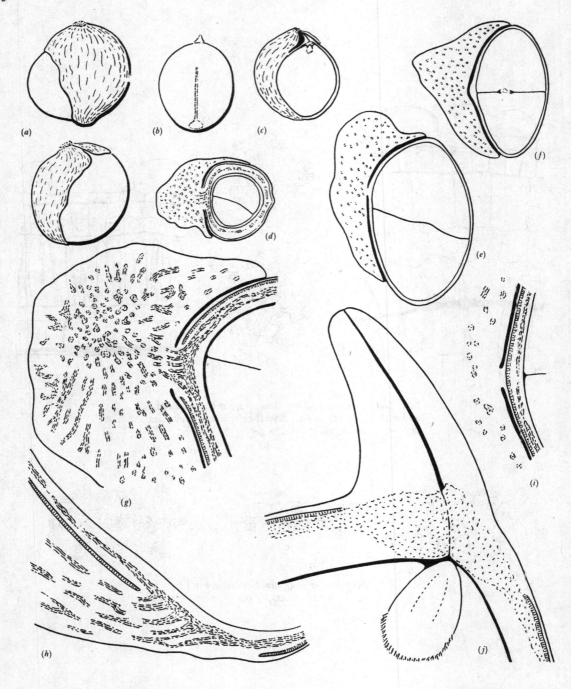

287 *Gonystylus sp.* (S. 29461): (*a*) seeds in side-view with the hilum upwards, × 1; (*b*) seed with the aril removed to show its attachment along the raphe, × 1; (*c*) seed in l.s., × 1; (*d*), (*e*), (*f*) seed in t.s. at the chalaza, in the middle, and below the micropyle, × 2; (*g*) chalaza in t.s., × 5; (*h*) chalaza in l.s., × 10; (*i*) raphe-attachment in the middle part of the seed, × 10; (*j*) micropyle in l.s., the brown tissue of the tegmen speckled, × 10.

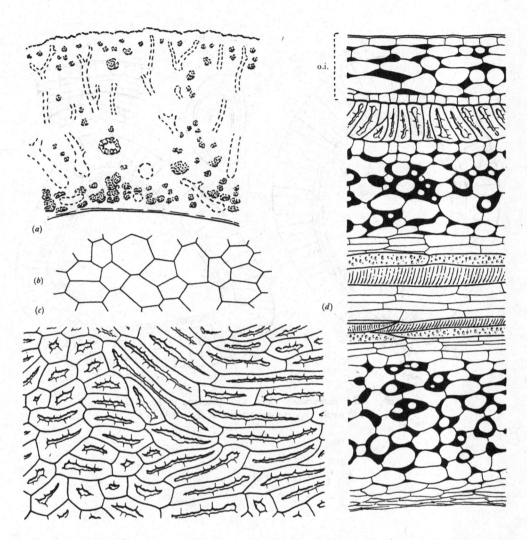

(a)

(b)

(c)

(d)

o.i.

288 *Gonystylus sp.* (S. 29461): (*a*) pericarp in t.s. with v.b. and groups of sclerotic cells, × 10; (*b*) exotesta and (*c*) exotegmen, in surface-view, × 225; (*d*) mature seed-coat in t.s., × 225.

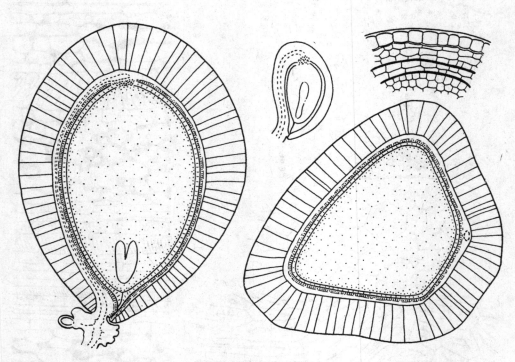

289 *Ribes sanguineum*. Seed with placenta in l.s. and
t.s., ×25; ovule, ×50; integuments of the ovule in t.s.,
×225.

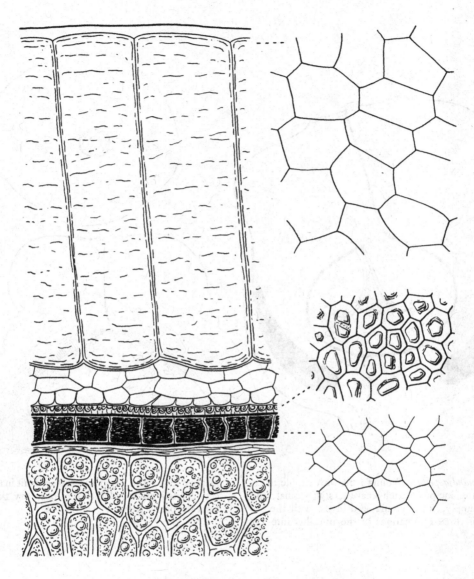

290 *Ribes sanguineum.* Seed-coats with endosperm in t.s., the endotegmic cells with brown contents, with epidermal facets, × 225.

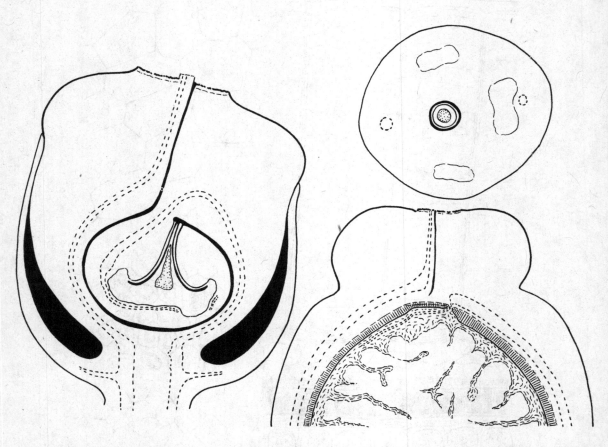

291 *Hernandia peltata*. Young fruit with cupule in l.s. to show the incipient pachychalaza, stylar canal, and vascular supply, × 12. Young seed in t.s. with the beginning of three ruminations of the nucellus into the massive testa, × 12. Apex of a mature fruit in l.s., with the cotyledon-tissue lobed into the loose dry tissue of the testa (or pachychalaza), × 5.

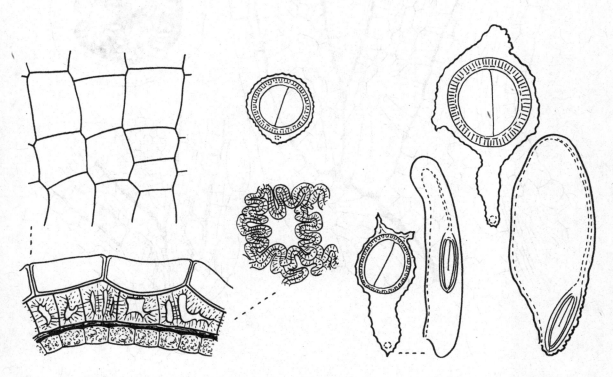

292 *Cratoxylon* with alate seeds (right); *C. arborescens* with central embryo, ×5; *C. cochinchinensis* with basal seed-body in l.s., ×10; in t.s., ×50. *Vismia sp.* (Corner 232; left), with seed in t.s., ×25; seed-coats in t.s., ×225.

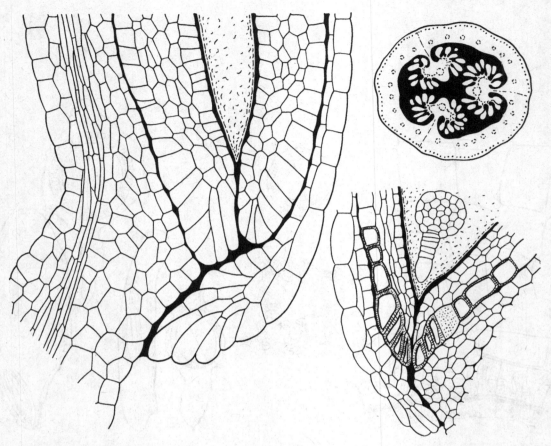

293 *Hypericum androsaemum.* Ovary in t.s., ×10. Micropyle of ovule (×500) and of the immature seed

(×225) with the lignification of the exotegmen proceeding from the endostome.

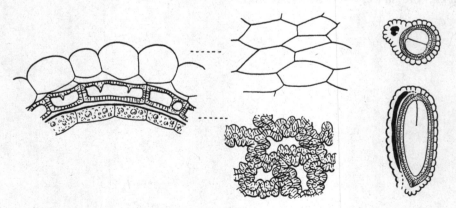

294 *Hypericum androsaemum.* Seed in l.s. and t.s., with the air-gap round the raphe v.b., ×25. Seed-coats in t.s. with endosperm, ×225.

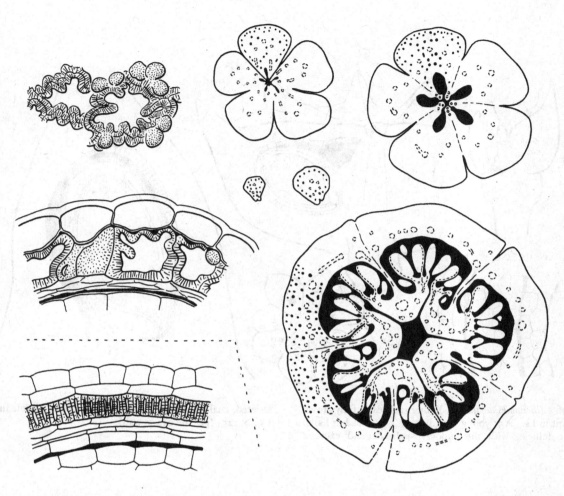

295 *Hypericum calycinum*. Young fruit in t.s., with the slime-canals of part of the pericarp shown in black, × 8. Style and top of the ovary in t.s., × 10. Seed-coats in t.s., with the lignified cells of the exotegmen in surface-view, × 225. *H. quadrangulare* (inset; after Guignard 1893), with seed-coats in t.s., × 225.

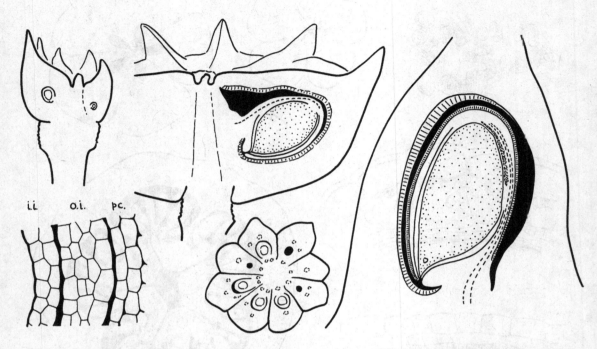

296 *Illicium kinabaluense*. Young and nearly fully grown fruit in l.s., × 5; young fruit in t.s., × 15. Seed in l.s. in the follicle, with the endocarp-palisade and exotesta striated, chalaza with tannin-cells, × 10. Wall of ovule in t.s., × 225; (pc. the ovary-wall).

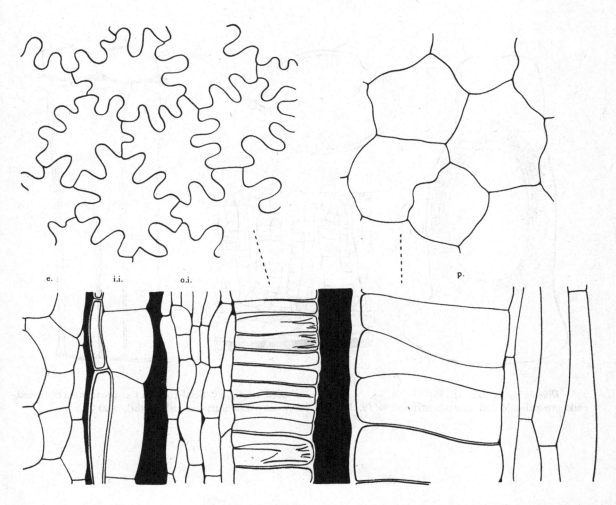

e.   i.i.   o.i.   p.

297 *Illicium kinabaluense.* Seed-coats of the nearly fully grown seed in t.s., with adjacent pericarp (p.) and endo- sperm, ×225; facets of exotestal and endocarp cells, ×500.

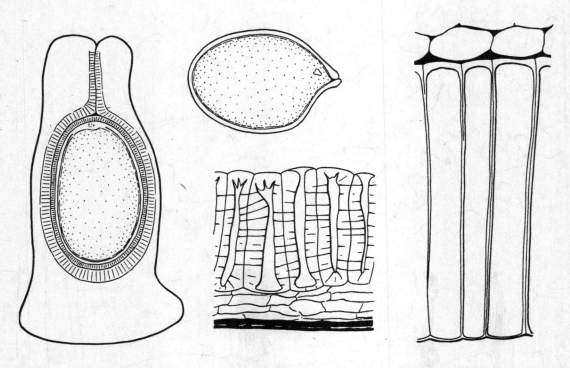

298 *Illicium sp.* (Malaya). Follicle in t.s. with the endocarp-palisade and exotesta striated, × 10. Mature seed in l.s., × 5. Seed-coats in t.s., the tegmen collapsed, × 225. Endocarp-palisade (right), × 225.

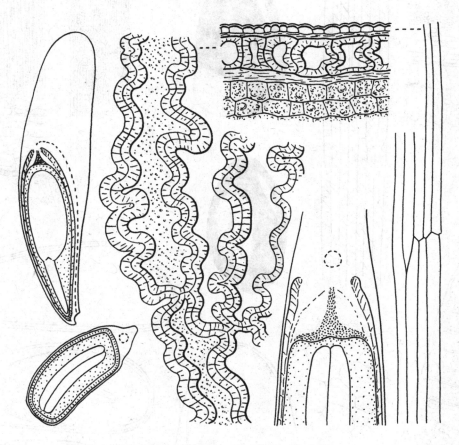

299 *Ixonanthes beccarii.* Seed in l.s., × 5; in t.s., × 10. Chalaza in transmedian l.s., × 25. Seed-coats in t.s. with endosperm, × 225.

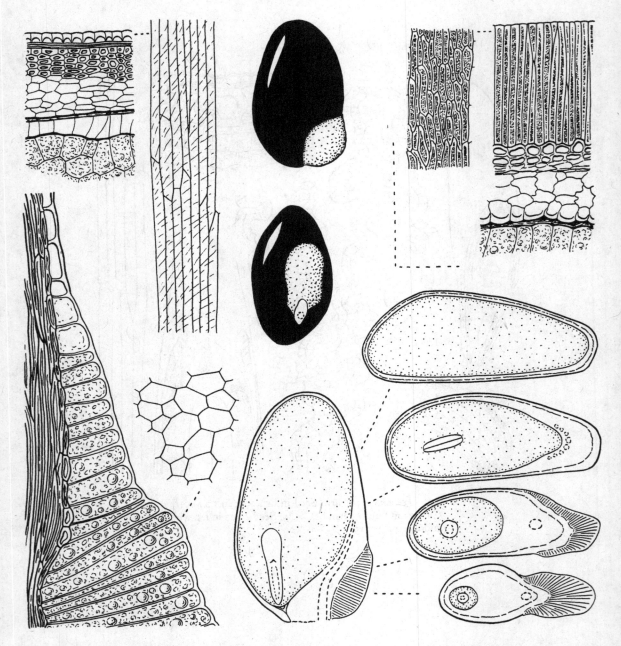

300 *Akebia lobata* and *Decaisnea insignis* (inset, upper right). *A. lobata*; seed in side-view and hilar view, ×5; seed in l.s. and t.s. at four levels, ×8; testa in t.s. with exotestal facets, and the junction of the aril-patch with the testa of the raphe, ×225. *D. insignis*; seed-coat in t.s. with exotestal facets, ×225.

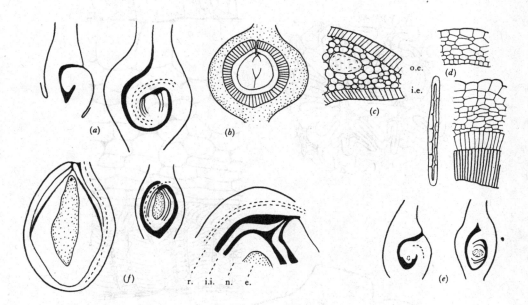

301 Lauraceae (after Sastri 1958*a*, 1962). (*a*) *Cassytha pubescens*, young ovary (left, ×65) and *C. filiformis*, mature ovary (×43). (*b*) *C. filiformis*, ripe fruit in l.s., the endocarp hatched, ×4. (*c*) *C. filiformis*, testa in l.s., with mucilage-canal, ×20. (*d*) *C. filiformis* young pericarp (above) in l.s., and fully grown pericarp with thick-walled endocarp-cells; i.h. with short, thin-walled palisade-cells, ×25; endocarp-cell, ×100. (*e*) *Cinnamomum zeylanicum*, young and fully grown ovary in l.s., mag. (*f*) *Litsea sebifera*, ovary and ovule in l.s. (left, ×18) and micropyle (right, ×100); r. raphe.

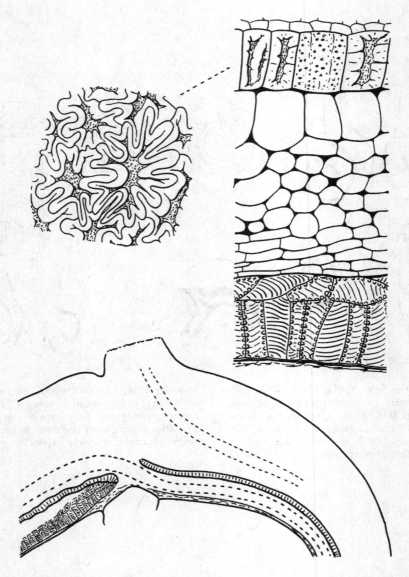

302 *Actinodaphne speciosa*. Stylar end of the fruit with endocarp (hatched), raphe v.b., tracheid tissue of the testa at the micropyle, radicle, and bases of the cotyle-dons, ×25. Testa in t.s. near the micropyle, with two inner layers of tracheids and the palisade-layer of the endocarp, ×225. Endocarp-facets, ×500.

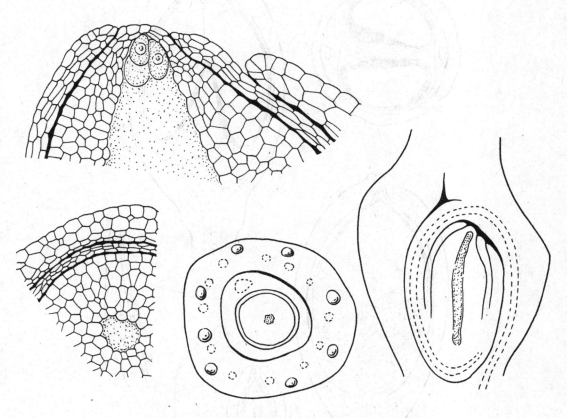

303 *Cinnamomum iners*. Ovary of open flower in l.s. and
t.s., × 30. Micropyle of ovule in l.s. and ovule-wall in
t.s., × 225.

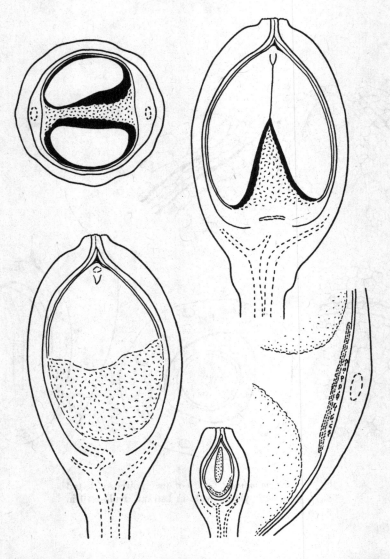

304 *Cryptocarya wightiana*. Mature and very young fruits in section within the perianth-tube, to show the ridge of nucellar tissue separating the divergent cotyle- dons, × 5. Perichalaza (lower right) in t.s. with hypostase and dilated v.b. inside the pericarp, × 15.

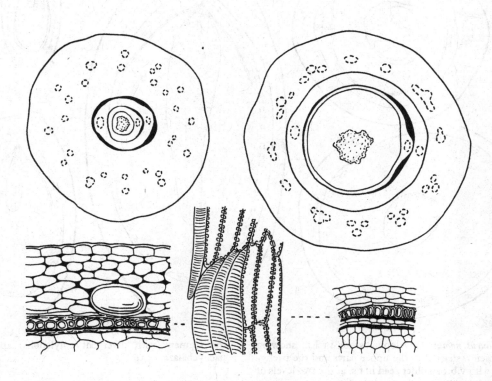

305 *Cryptocarya wightiana.* Young fruit in t.s. near the apex of the seed and at the middle, with concentric perianth-tube, pericarp, seed-coat, nucellus and endosperm, ×25. Mature seed-coat (lower left) with pericarp in t.s., and young seed-coat (lower right) with nucellus, tegmen and testa, ×225; tracheids of the endotesta, ×500.

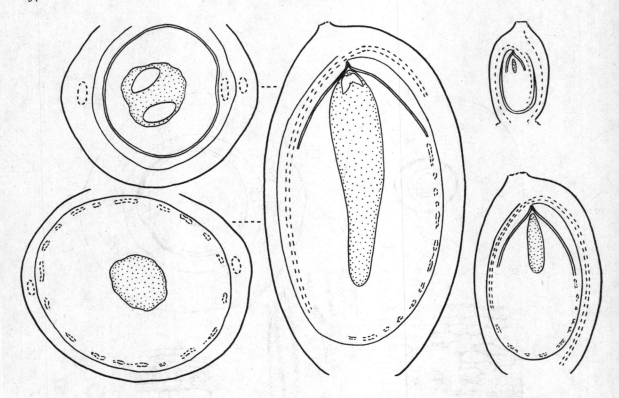

306 *Laurus nobilis*. Developing seeds in l.s., showing the tegmen restricted to the upper parts and the pachychalaza with v.b.; an older seed in t.s. at the two levels of the tegmen (with divergent cotyledons) and of the pachychalaza, × 10.

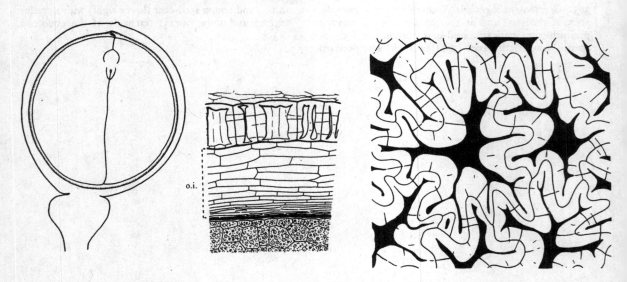

307 *Laurus nobilis*. Fruit in l.s., showing the seed enclosed in the woody endocarp (hatched), × 5. Mature seed-coat with endocarp and cotyledon-tissue, × 225. Endocarp facets, × 500.

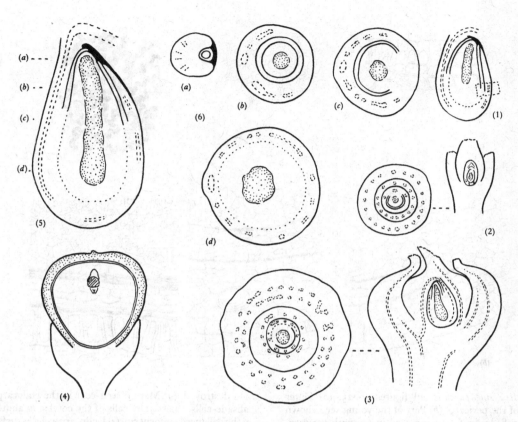

308 *Litsea singapurensis.* (1) Young seed soon after fertilization in l.s. (the rectangle enlarged in Fig. 309), ×30. (2) Young fruit soon after fertilization in l.s., ×7; in t.s., ×14. (3) Young fruit *c.* 3 weeks old in l.s. and t.s., ×7. (4) Mature fruit in l.s., one cotyledon removed, the pericarp stippled, ×7. (5) Young seed, as in (3), ×30. (6) Young seed, as in (3), in t.s. at successive levels, ×30. (In (1) and (5), the separation between the large-celled nucellus and the small-celled chalaza shown by a dotted line.)

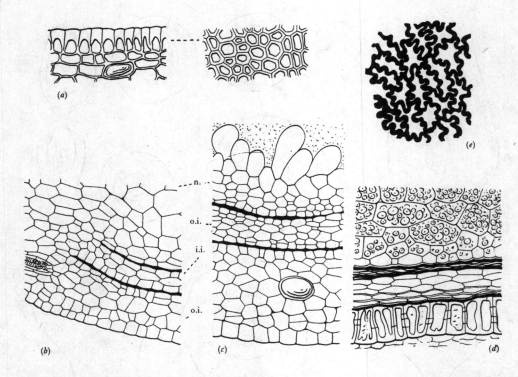

309 *Litsea singapurensis*, all figures ×225. (*a*) Outer surface of the pericarp. (*b*) Part of the young seed shown in Fig. 308, in l.s. (*c*) Young seed in t.s. with pericarp, nucellar cells (n.) enlarging before absorption, the teg-men destroyed. (*d*) Mature seed-coat in the endocarp of palisade-cells, the starchy cells of the cotyledon abutting on the collapsed endosperm. (*e*) Endocarp-cells in surface-view.

310   *Couroupita guyanensis*. Seed with coma of hairs in
t.s. and l.s. (embryo intact), and embryo in l.s., × 5.

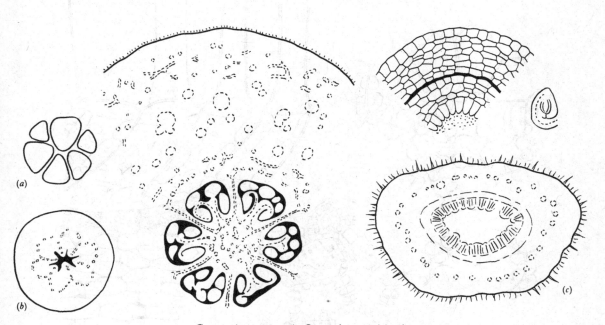

311   *Couroupita guyanensis*. Ovary in t.s.: (*a*) stigmata;
(*b*) apex of ovary, × 10. Ovule, × 25; ovule-wall, × 225.
(*c*) Pedicel of flower in t.s. with cortical v.b., × 10.

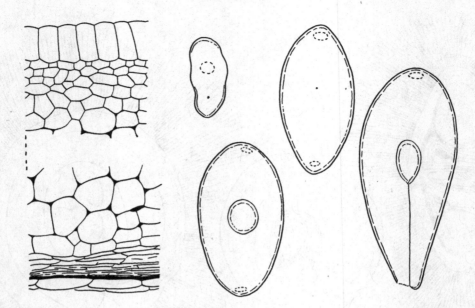

312 *Couroupita guyanensis.* Young seed in transmedian l.s. (right) and in t.s. at three levels from the nucellar region towards the micropyle, × 10. Seed-coats of the young seed in t.s., the inner layer of the testa collapsing, the tegmen 2 cells thick, × 225.

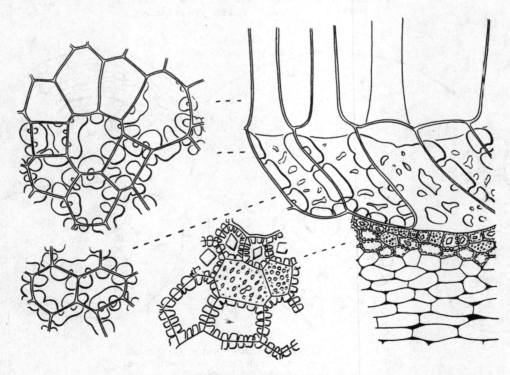

313 *Couroupita guyanensis.* Outer part of the mature testa in l.s., with cell-details in tangential section, × 225; o.h. in tangential section, × 500.

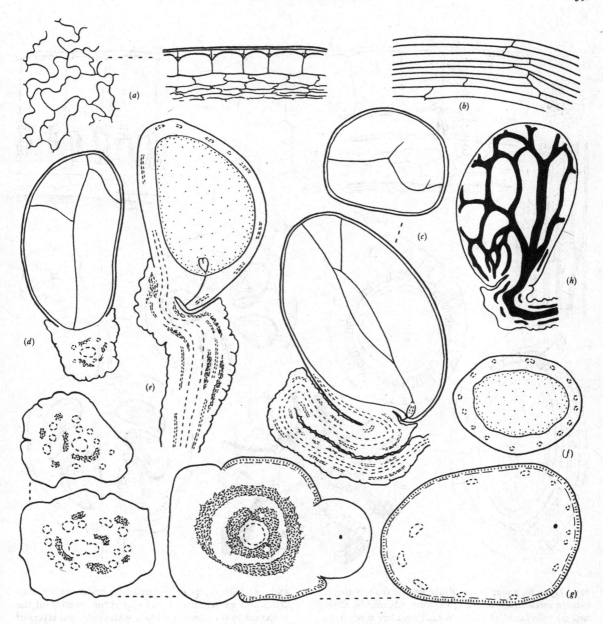

314 *Gustavia sp.* (*a*) Seed-coat in t.s., reduced to the exotesta and crushed mesophyll, × 225; facets of the exotesta, × 110. (*b*) Epidermal cells of the funicle in surface-view, × 110. (*c*) Mature seed and funicle in l.s., and the seed in t.s., × 3. (*d*) Mature seed in transmedian l.s. to show the slight aril-lobes of the funicle, × 3. (*e*) Immature seed in l.s., the funicle not yet folded, × 3. (*f*) Immature seed in t.s., × 3. (*g*) Funicle and base of the seed in t.s. from the base of the funicle to the micropyle (4 sections), the sclerotic tissue stippled, the exotesta striated, × 8. (*h*) Vascular bundles on one side of the seed, × 3.

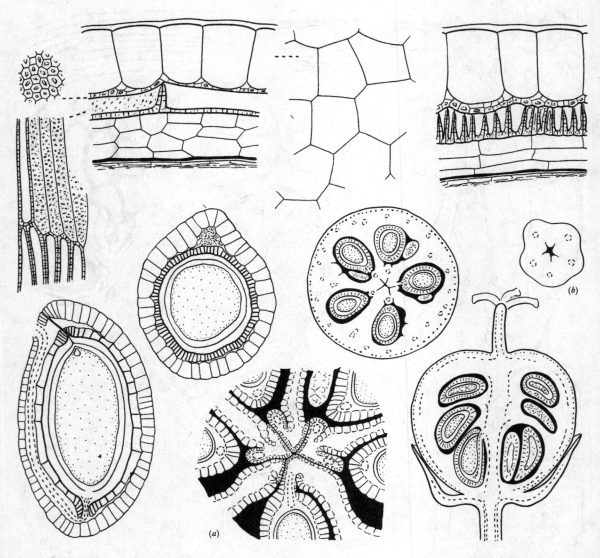

315 *Gynotroches axillaris*. Fully grown fruit with immature seeds in l.s. and t.s., showing v.b. and hypodermal oil-cells (as black dots), × 12. Immature seeds in l.s. and t.s., showing the exotestal palisade, exotegmic fibres, nucellus, endosperm, and v.b., × 50. Seed-coats in l.s. and t.s., with facets of the exotesta, endotesta (crystal-cells) and exotegmen, × 225. (a) Axile meeting of the placentas in the immature fruit, with epidermal layers of brown mucilage-cells, × 25. (b) Style in t.s., × 50.

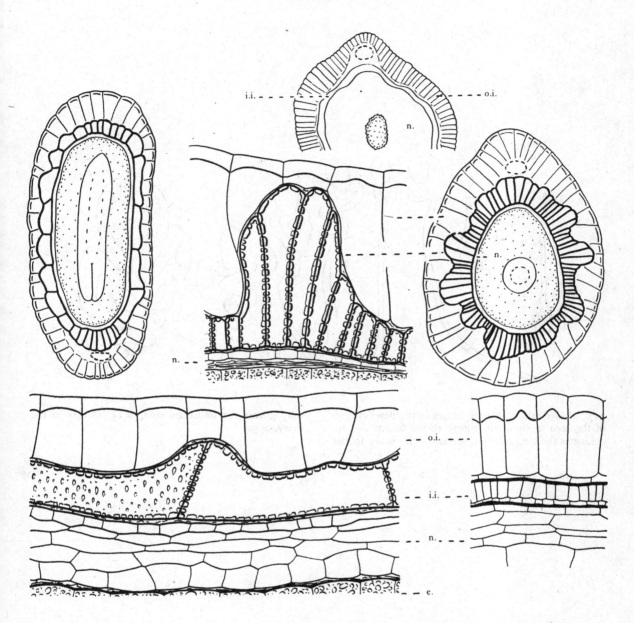

316 *Pellacalyx axillaris*. Mature seed in transmedian l.s. (×40, left) and in t.s. (×60, right). Immature seed with thick nucellus in t.s. (×60, upper centre). Mature seed-coat in t.s. and l.s., with the testa and tegmen reduced to single layers of cells, and an immature seed-coat with 2-layered testa and tegmen, ×225.

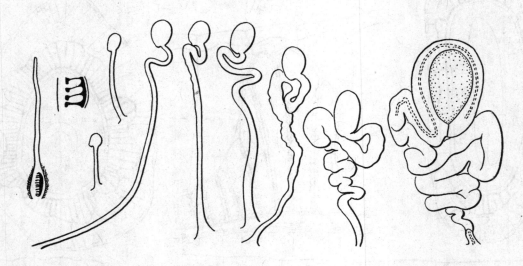

317 *Acacia auriculaeformis*. Stages in the development of the seed to show the growth of the funicle and its coiling as the seed enlarges; section of the ovary to the fully grown but immature seed, × 15. Ovules in the ovary, × 50.

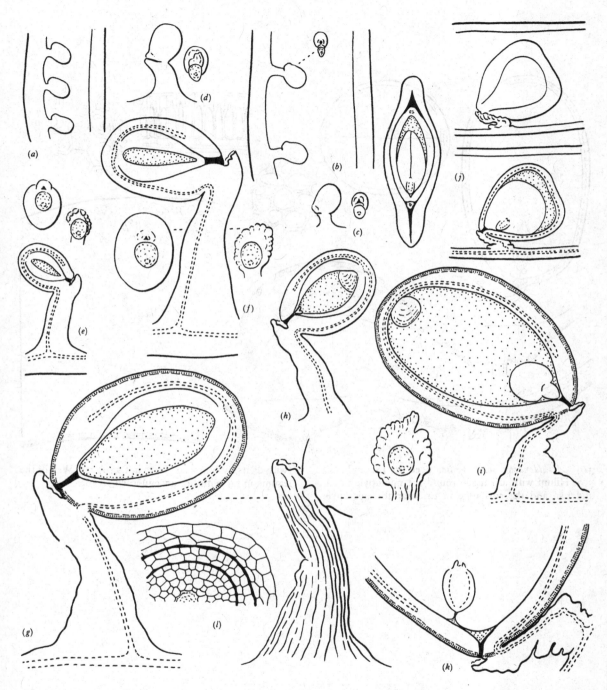

318 *Adenanthera pavonina*. (*a*)–(*g*), Stages in the development of the seed from the ovule at anthesis (*a*), showing at several stages the head of the funicle with vestigial aril and the hilum of the detached young seed, × 15. (*h*), (*i*) Immature seed with developing embryo, endosperm with watery space at the chalazal end, and, in (*h*), the remains of the nucellus, in l.s., × 6. (*j*) A fully grown but immature seed with remains of the endosperm, in side-view, l.s. and t.s., × 2. (*k*) The micropylar end of a fully grown seed in l.s., with compressed funicle and vestigial aril, × 10. (*l*) Ovule in t.s. at anthesis, × 225.

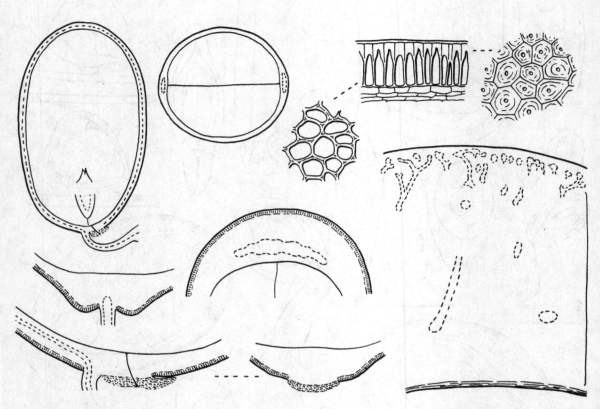

319 *Archidendron solomonense*. Seed in l.s. and t.s., × 3. Hilum with aril-tissue round the micropyle in l.s. and t.s., and the antiraphe in t.s. near the micropyle (with dilated v.b.), × 10. Testa in t.s., × 225. Wall of the legume in t.s. with fibrous endocarp, × 10.

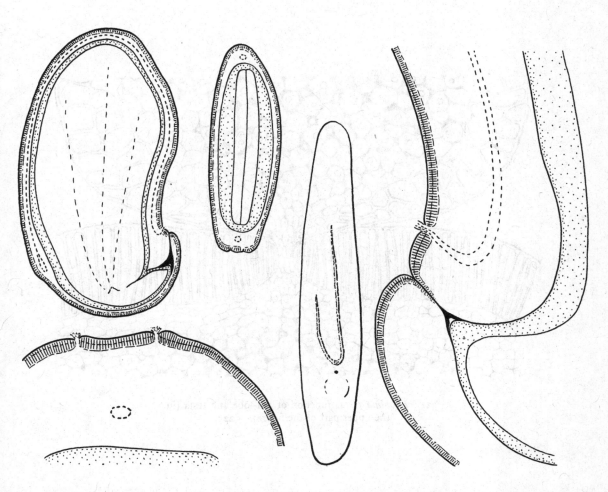

320 *Barklya syringifolia.* Seed in median l.s., × 5. Seed in t.s. and in hilar view with micropyle and crescentic scar of the funicular attachment (as in *Bauhinia*), × 8.

Hilum in l.s. with funicle-entry and in t.s. with the two scars of the crescentic attachment, × 25.

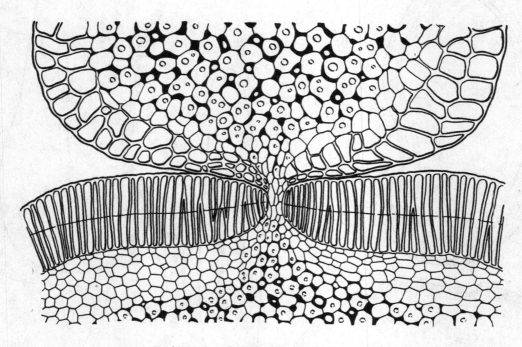

321   *Bauhinia picta.* Junction of aril-lobe and testa (in
the upper part of the figure), ×225.

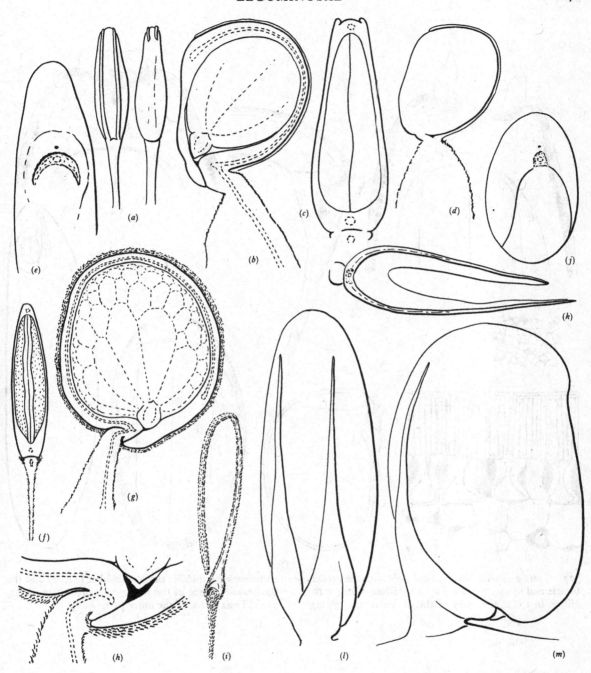

322 Seeds of *Bauhinia*.

B. kockiana. (*a*) Seeds in raphe- and antiraphe-view with long aril-lobes, × 5; (*b*) immature seed in l.s., × 6; (*c*) mature exalbuminous seed in transmedian l.s., × 6; (*d*) seed and funicle in side-view, × 5.

B. integrifolia. (*e*) Hilum without aril-lobes, × 5; (*f*) seed and funicle in transmedian l.s., × 2; (*g*) seed in l.s., × 2; (*h*) micropylar end of the seed with funicle and peripheral band of endocarp-hairs, × 5; (*i*) funicle with the hoop of endocarp-hairs detached from the seed, × 2.

B. acuminata. (*j*) Seed in hilar view with lines of detachment of the aril-lobes, × 6; (*k*) funicle with aril-lobes detached from the seed, × 6; (*l*), (*m*) seed in raphe-view and side-view, × 6.

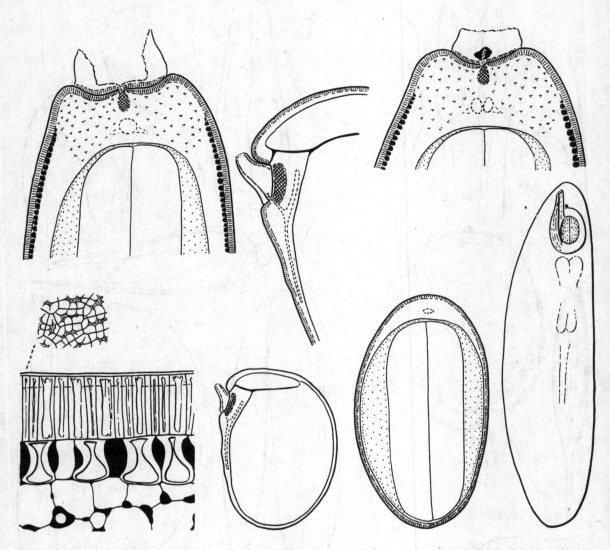

323 *Cadia purpurea*. Seed in hilar view with the broken funicle and in t.s., × 10; in l.s., × 5. Hilum in l.s., × 10. Hilum in t.s. at the entry of the funicular v.b. (right), and across the middle, the tracheid-bar hatched and the thick-walled tissue of the hilum with substellate marks, × 25. Testa in t.s. of the outer part, × 225.

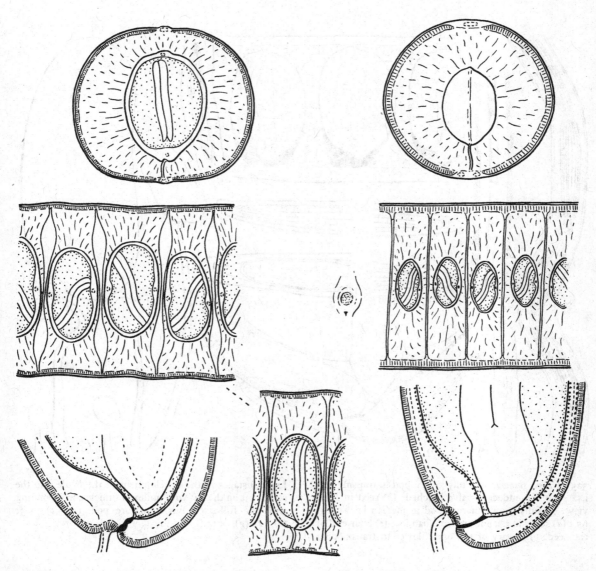

324 *Cassia javanica* (left) and *C. fistula* (right), *C. java-nica*; pod in t.s., transmedian l.s. and median l.s., ×3. *C. fistula*; pod in t.s. and transmedian l.s., ×2; hilum, ×30. Hilar ends of the seeds in l.s., ×30. Endocarp-valves hatched.

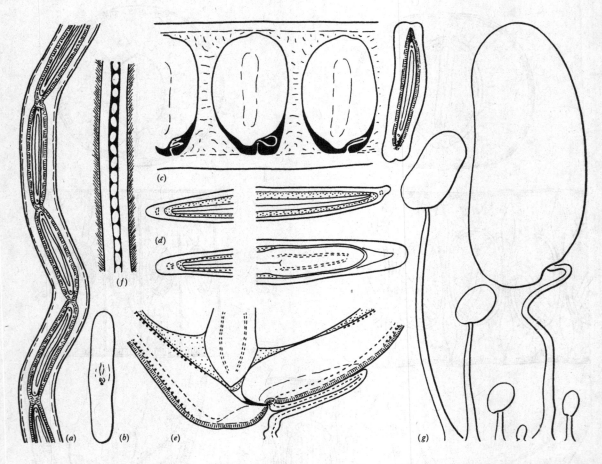

325 *Cassia siamea.* (*a*) Fully grown pod in transmedian l.s., with the endocarp-valve hatched; (*b*) seed in hilar view; (*c*) fully grown mature pod in median l.s. and in t.s.; (*d*) seeds in t.s. and transmedian l.s.; (*e*) hilar end of the seed; (*f*) ovary of the open flower in transmedian l.s.; (*g*) stages in the development of the ovule into the seed, that on the left with unflexed funicle corresponding with the full-sized but immature pod. (*a*), (*c*), ×3; (*b*), (*d*)–(*g*), ×15.

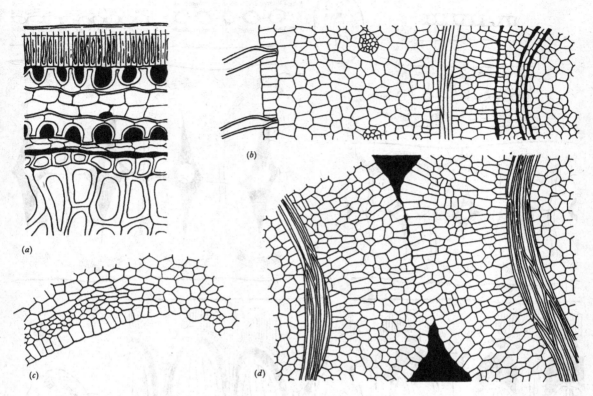

326 *Cassia siamea*. (*a*) Testa in t.s. with endosperm and (?) a single layer of perisperm, the tegmen effete; (*b*) part of the very young pod in transmedian l.s., with incipient endocarp-valve and endocarp-pith, and part of the enlarging ovule; (*c*) the embryonic endocarp of the carpel of the open flower in t.s.; (*d*) the development of the endocarp as seen in transmedian l.s., to form the septum of the pod; × 225.

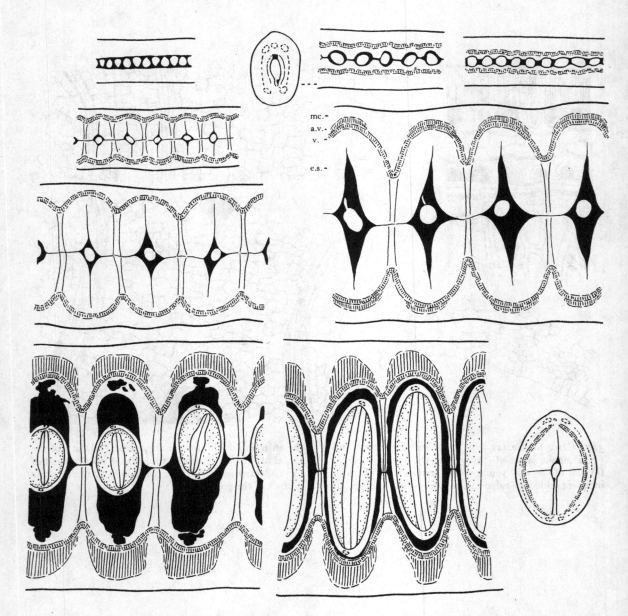

327 *Cassia spectabilis*. Ovary and developing pod in transmedian l.s. and in t.s., showing the development of the endocarp-valve (v.), the accessory endocarp-valve (a.v.), the endocarp septa (e.s.), the mesocarp (mc.) and the seed; uppermost figures ×15, the rest ×7.

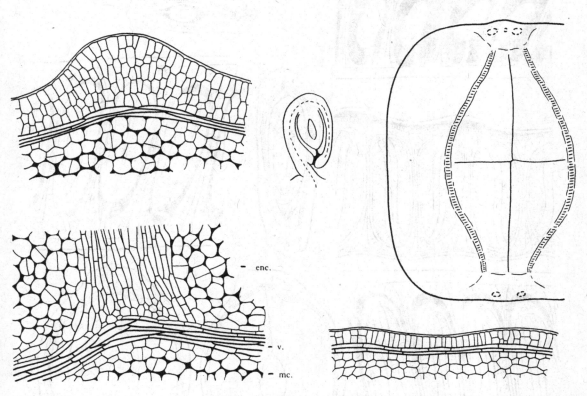

328 *Cassia spectabilis*. Ovule, × 50. Fruit in t.s. with septum in surface-view, × 7. Sections of the developing endocarp (enc.) showing its origin from i.e. (pericarp), the origin of the endocarp-valve (v.) from i.h. (pericarp), and the transversely elongate cells of the fruit-septum in the endocarp; mc. mesocarp; × 225.

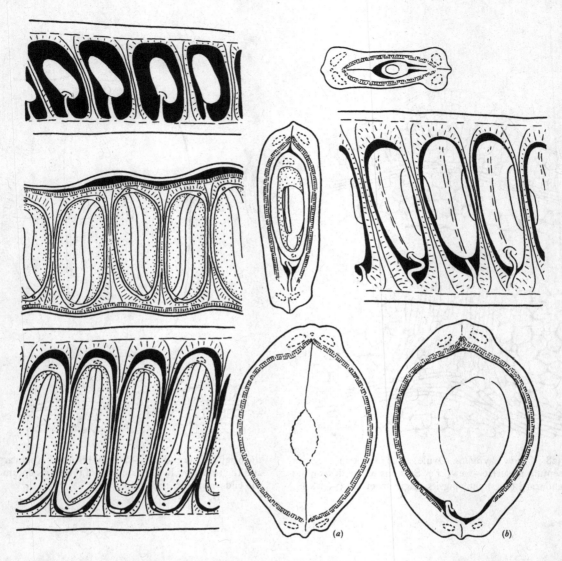

329 *Cassia occidentalis.* Young and nearly mature pods in l.s. and t.s., with the endocarp-valve hatched; (*a*) full-sized pod cut at the septum to show the central hole and (*b*) the mature seed with pleurogram; ×6.

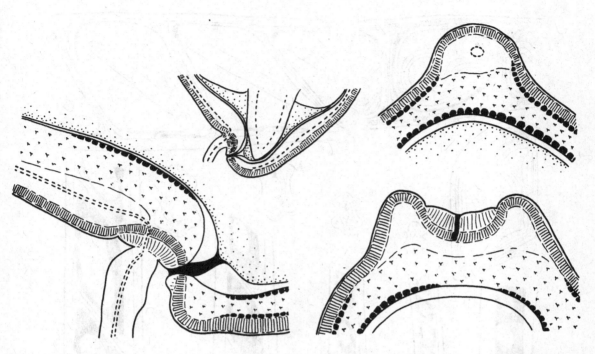

330 *Cassia occidentalis*. Hilar end of the seed in l.s., × 15. Details of the hilar end in l.s. and t.s. and the beginning of the raphe in t.s., with the hourglass–cells and the aerenchymatous mesophyll, × 50.

331 *Cassia tora*. Mature pods in median and trans-median l.s., with the endocarp-valves hatched, and de-veloping pods in median l.s. and in t.s., ×6; (*a*) ovary of the open flower in l.s., ×15.

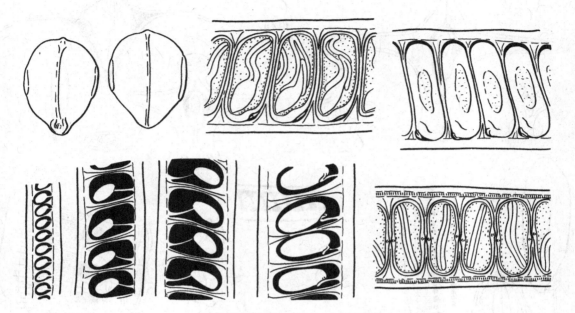

332 *Cassia hirsuta*. Seeds in raphe- and antiraphe-view, mature pods in median and transmedian l.s., and developing pods in median l.s., (the hairs omitted), ×6.

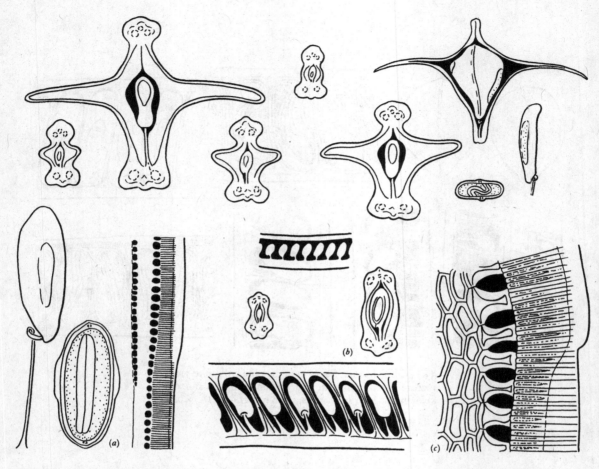

333 *Cassia alata* (upper figures) showing in t.s. the development of the ovary into the pod, × 6; seed in side-view and in t.s., × 2. *C. multijuga* (*a*), the mature seed in side-view and t.s., × 6, with a diagram of the structure of the testa in t.s. at the edge of the pleurogram, × 75. *C. occidentalis* (*b*), young pods in l.s. and t.s., the older at the time of rotation of the seeds, × 6 (cf. Fig. 329). *C. hirsuta* (*c*), the edge of the pleurogram in t.s., × 225.

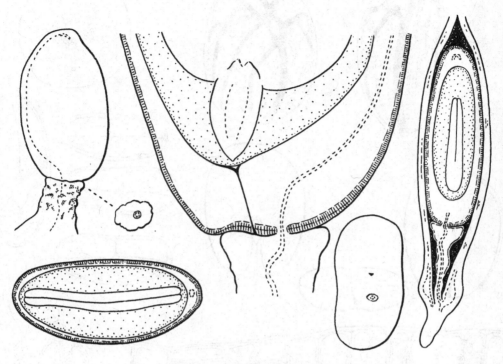

334 *Cercis siliquastrum*. Immature seed in side-view, with broad funicle, the direction of the v.b. indicated by a broken line, ×8; the funicle-head detached to show the minute central area of attachment to the seed, ×10. Immature seed in transmedian l.s. in the pod, and a mature seed in t.s., ×10. Base of the seed with micropyle and hilum, ×25. Hilar end of seed in l.s., ×25.

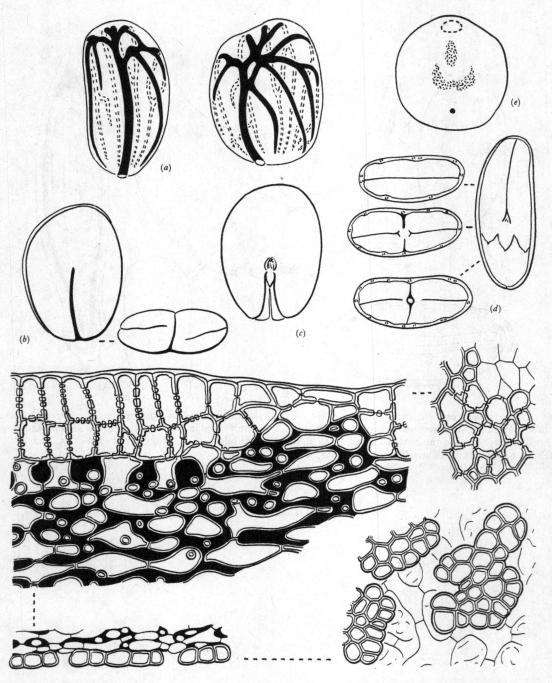

335 *Cordyla africana*. (a) Vasculature of the seed, × 1; (b) embryo in side-view and in basal view, × 1; (c) embryo in median l.s., × 1; (d) seed in l.s. and t.s., × 1; (e) micropylar region in t.s. to show aerenchyma (speckled) and absence of tracheid bar, × 5. Testa in t.s., showing a thin-walled area, and the endotesta, × 225.

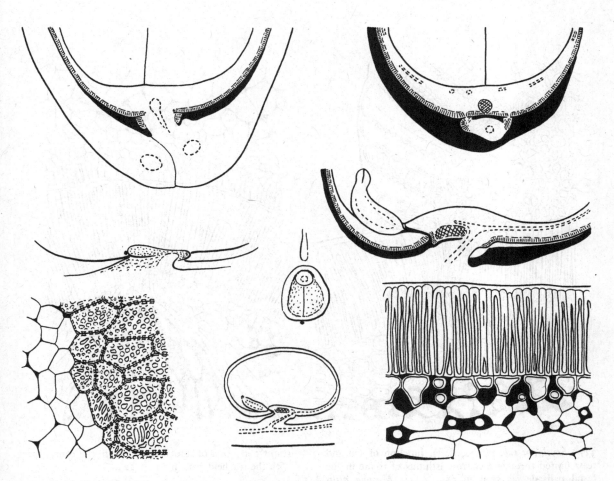

336 *Erythrina subumbrans*. Seed in l.s., ×2. Hilum in side-view with the horny part of the funicle and the rim-aril stippled, and the hilum in surface-view with the funicle detached (except for the rim-aril), ×6. Hilar side of the seed in l.s. and t.s. at the entry of the v.b. (upper left) and at the tracheid bar (upper right), ×6. Surface of the testa in t.s. and the cells of the tracheid-bar, ×225.

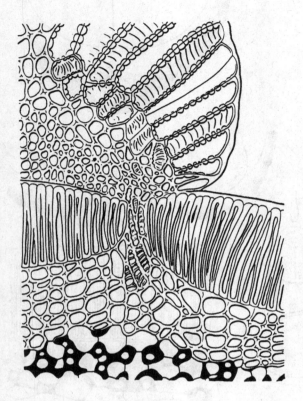 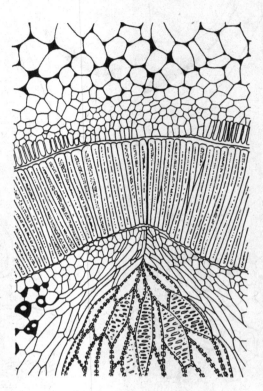

337 *Erythrina subumbrans*, (left); junction of aril and testa united through a narrow isthmus of tissue in the testal palisade, as seen in t.s., ×225. *Mucuna utilis* (right); junction of funicle and testa in the central part with the tracheid-bar, in t.s., ×225.

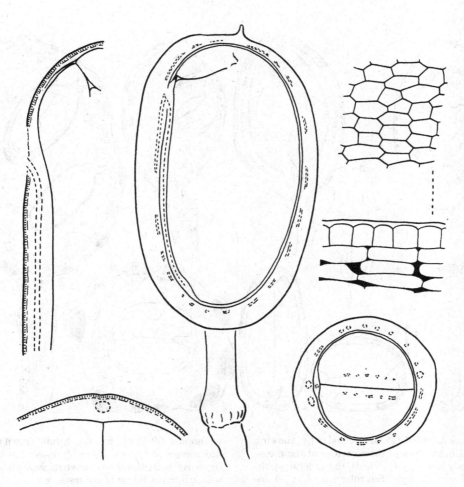

338 *Euchresta japonica*. Pod with mature seed in l.s. and t.s., × 5. Hilum in l.s. and raphe in t.s., × 10. Outer part of the testa, × 225.

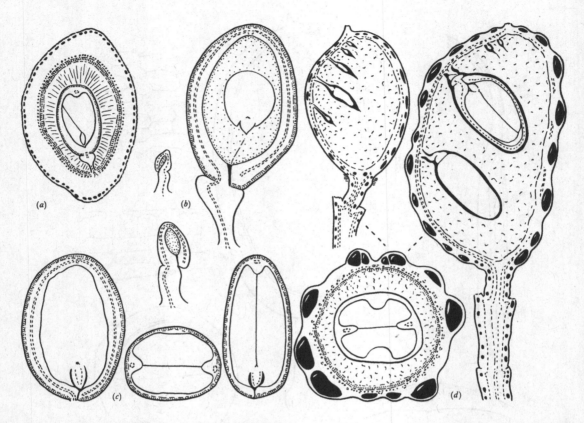

339   *Hymenaea courbaril.* (*a*) Mature pod in t.s., showing the outer and inner rows of resin-cavities of the mesocarp, the woody endocarp (hatched), the endocarp-pith, and the seed with vestigial funicular aril, × 1; (*b*) developing seeds, × 3; (*c*) mature seeds, × 1½. *Trachylobium verrucosum.* (*d*) Young and nearly fully grown pod in l.s., and mature pod in t.s., as in *Hymenaea* but with much larger resin-cavities in the outer mesocarp and the seed with 4 internal ridges of the testa, × 2.

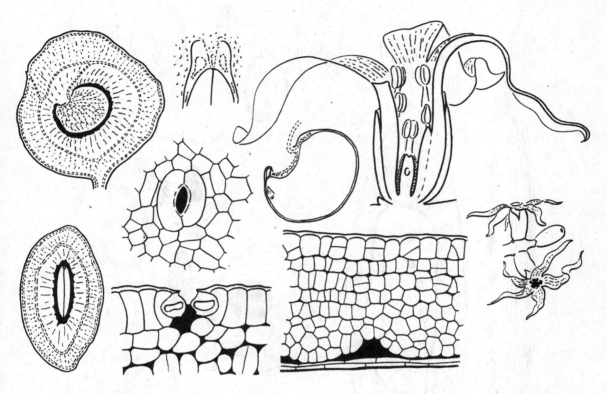

340 *Inocarpus edulis.* Fruit in l.s. and t.s. with double vascular supply connected by radiating fibro-vascular bundles in the mesocarp, with air-space between the endocarp-pith and the seed, $\times \frac{1}{2}$. Seed in l.s. to show the recurrent v.b., $\times 1$; the hilum in t.s., $\times 2$. Testa in t.s., $\times 225$, and stomata of the testa, $\times 500$. Flowers, $\times 2$, and a flower in l.s. showing the epipetalous stamens and radial symmetry, $\times 7$.

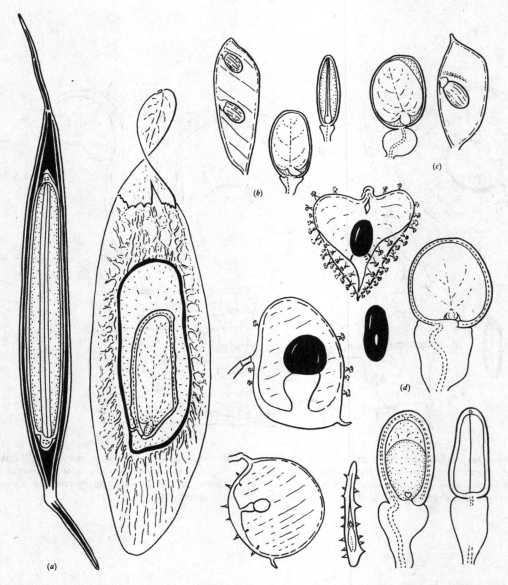

341 *Koompassia malaccensis* (a): mature pod in l.s., ×1; in t.s., ×5. *Petalostylis labicheoides* (b): open pod, ×1½; seed in median and transmedian l.s., ×4. *Labichea rupestris* (c): pod, ×1½; seed, ×4. *Sindora wallichii* (d): mature pods (one split open) with resin-capped spines and swollen red funicles, with immature pods in l.s. and t.s., ×½; mature seed in median and transmedian l.s., ×1½; immature seed with rudimentary embryo in the endosperm, ×3.

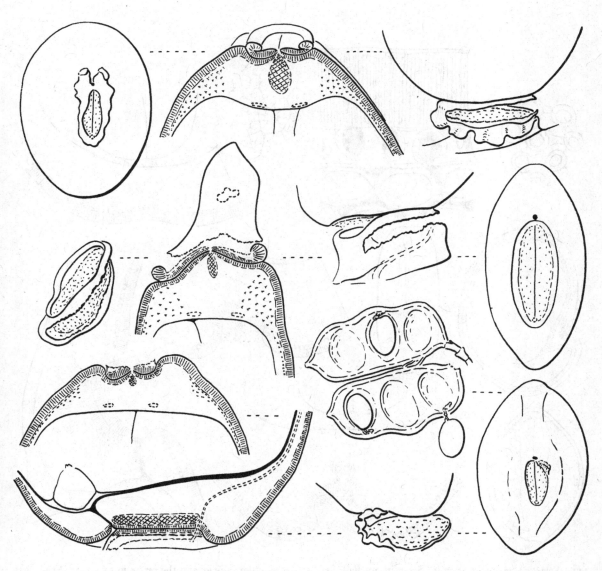

342 *Ormosia semicastrata* (upper row), *O. henryi* (middle row), *O. bancana* (lower rows, apparently without trace of aril). Seeds and funicles, × 6. Hilar ends of the seeds in section, × 15.

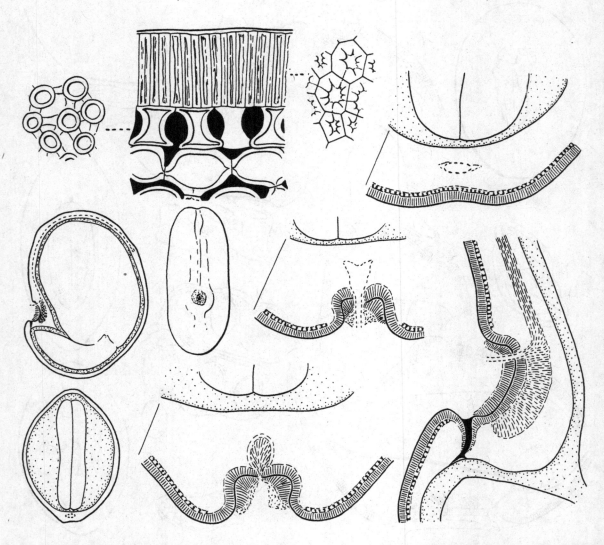

343 *Swartzia madagascariensis.* Seed in median l.s., t.s., and hilar view, × 5. Hilum with tracheid-bar and micropyle in l.s., and in t.s. across the tracheid-bar and across the funicle, and the raphe in t.s., × 25. Seed-coat with palisade and hourglass-cells, × 225.

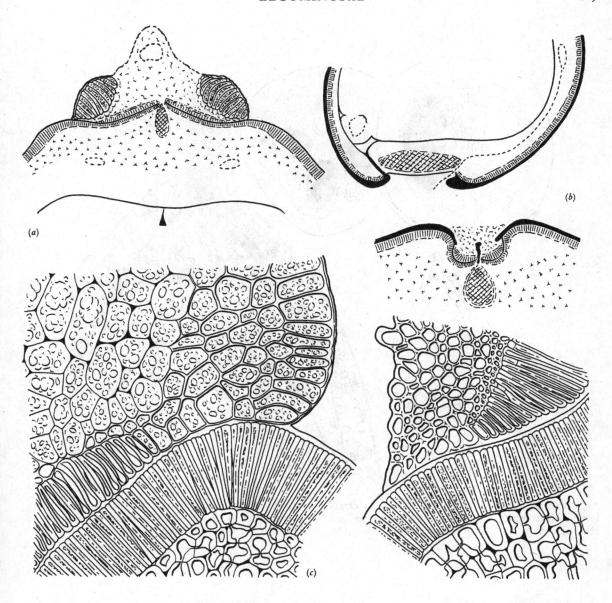

344 *Vigna sinensis* (*a*): the papilionaceous hilum with attached head of funicle in t.s., to show the large cells of the vestigial aril with brownish contents, × 30. *Sophora tomentosa* (*b*): with the hilar end of the seed in l.s., × 6, and t.s., × 15, and the edge of the vestigial aril, × 225. *Tephrosia candida* (*c*): the edge of the vestigial aril, the cells filled with starch, × 225.

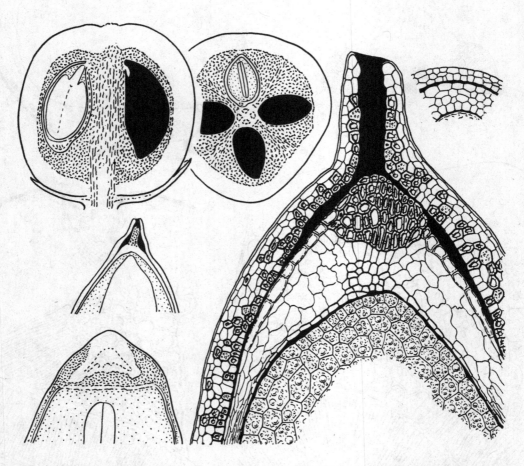

345 *Hugonia mystax*. Fruit in t.s. and l.s. (3 empty loculi), with woody endocarp, ×5. Micropylar and chalazal ends of the seed in transmedian l.s., the endo-stome and the testa around the chalaza woody, ×25. Ovule-wall in t.s., ×225. Micropylar region in trans-median l.s., with endosperm, ×225.

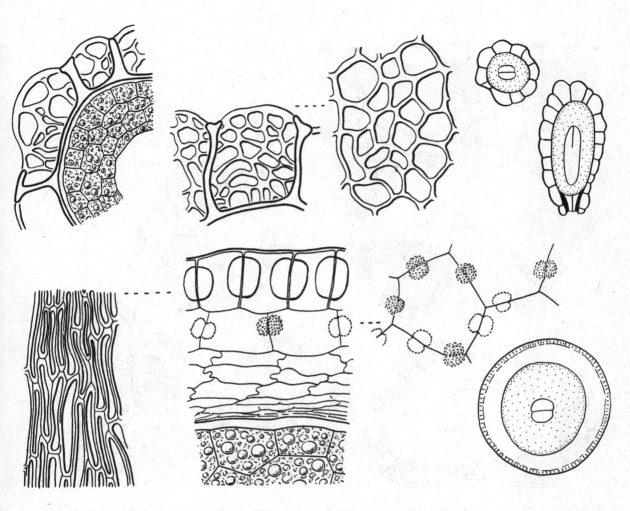

346 *Cajophora sp.* (upper row): seed in l.s. and t.s., ×25; section of the seed with the large cells at the chalazal end, ×120; epidermal facets, ×50. *Blumen-* *bachia insignis* (lower row); seed in t.s., ×25; section of the seed-coat, ×225; epidermal facets, ×50; hypodermal cell with callus-patches in tangential view, ×225.

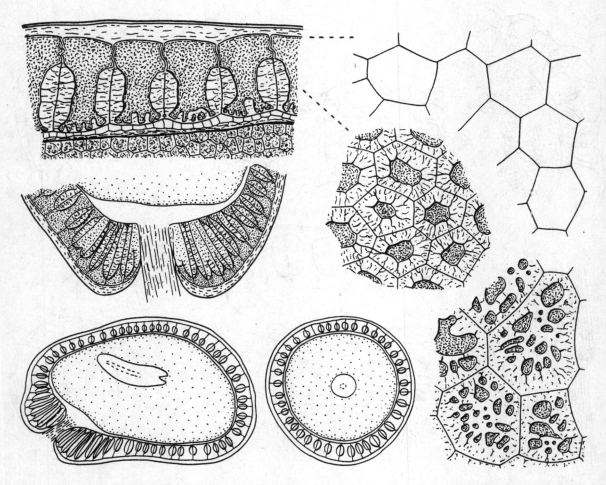

347 *Fagraea ceilanica.* Seed in l.s. and t.s., covered by the gelatinous pellicle, ×25. Chalaza in l.s., ×50. Testa in t.s., with gelatinous pellicle from the outer cell-walls of the epidermis separating the cuticle, epidermal cells with dark brown contents, crushed mesophyll, and endosperm, ×225. Facets of the epidermal cells and sections of them across the middle region, ×225; inner facets, ×500.

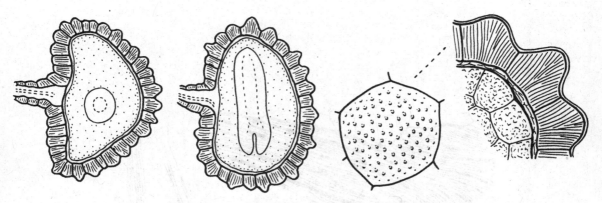

348 *Geniostoma randianum.* Seed in l.s. and t.s., × 50.
Part of the seed-coat and endosperm, × 225; epidermal
cells in optical section, × 500.

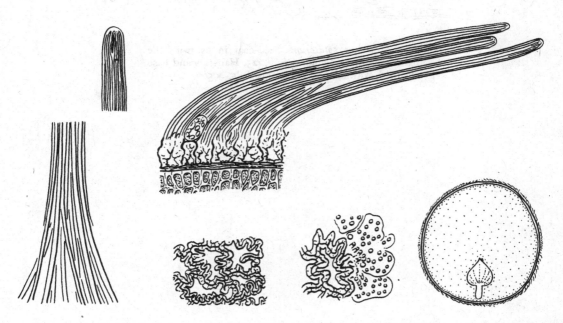

349 *Strychnos nux-vomica.* Seed in l.s., × 1½. Testa and
endosperm in section, × 120. Hair (left) with broken base
showing the lignin-strands, and with intact tip, × 225.

Undulate facets of the outer epidermal cells (lower
centre), that on the right with lignin-strands aggregating
into the hairs, × 225.

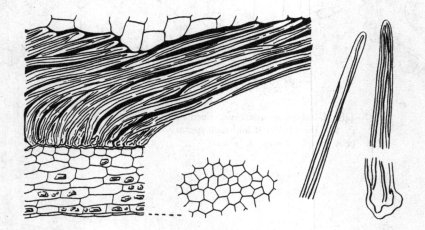

350 *Strychnos potatorum*. Seed-coat in l.s. with the
hairs and adjacent endocarp, × 225. Hair-tips and base
enlarged, with internal thickening rods, × 500.

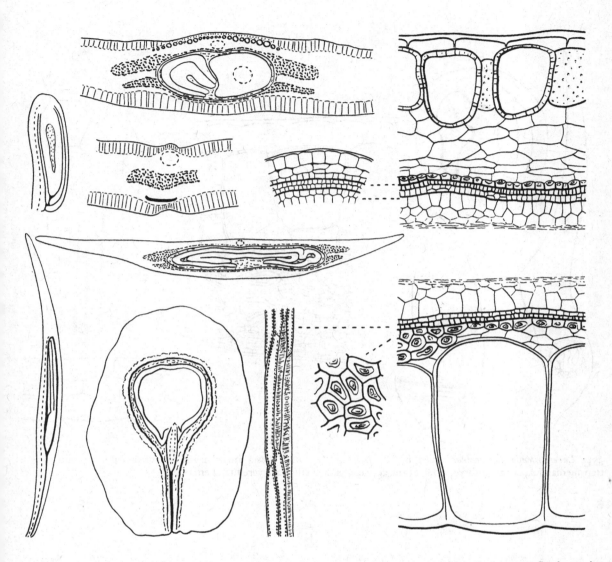

351 *Lafoensia densiflora*. Seed in flat view with the testa cut away to show the embryo and long micropyle, ×4; in l.s., ×5; in t.s., ×12; micropylar region and hypocotylar region with v.b., showing the exotestal palisade, the hypodermal sclerotic cells on the raphe-side, and the mesotestal sclerotic flanges, ×25. Ovule, ×25; ovular integuments and nucellus, ×225. Seed-coat in t.s., with sclerotic hypodermal cells on the raphe-side (above) and on the antiraphe-side (below), at the junction of the mesotestal flange, ×225. Facets of endotesta (×225) and of the tracheidal exotegmen (×500).

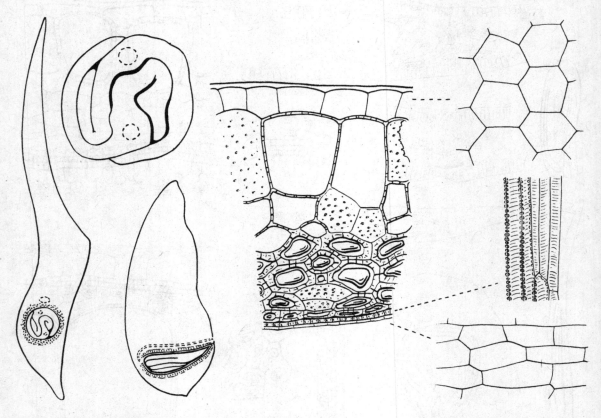

352 *Lagerstroemia flos-reginae.* Seed in l.s., × 5; in transmedian l.s., × 10. Embryo in t.s. of the cotyledons, × 25. Testa in t.s. on the side of the embryo, with very thin 2-layered tegmen, × 225.

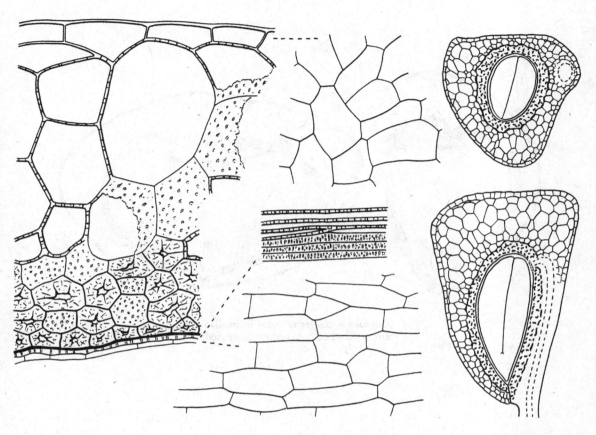

353 *Lawsonia inermis*. Seed in l.s. and t.s., with large-celled outer mesotesta and small-celled, densely sclerotic, inner mesotesta, ×25. Seed-coat in t.s., with facets of exotesta, exotegmen and endotegmen, ×225.

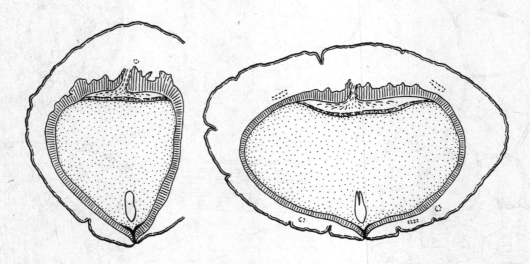

354 *Magnolia soulangeana*. Seed in median and
transmedian l.s., ×8; endotesta striated.

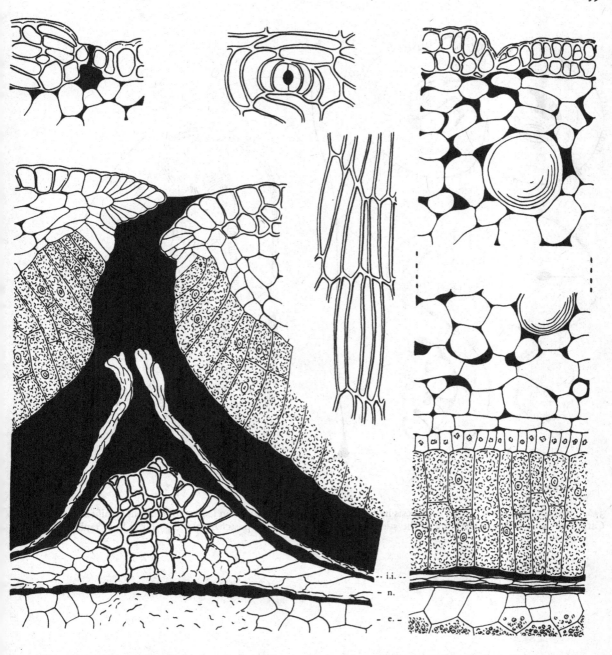

355 *Magnolia soulangeana.* Micropyle of the fully grown but immature seed in transmedian l.s., × 225; endotesta with internal fibrils in the cells, tegmen collapsed, nucellar apex thick-walled. Seed-coats in t.s., with stomata and outer epidermal cells in surface-view, × 225.

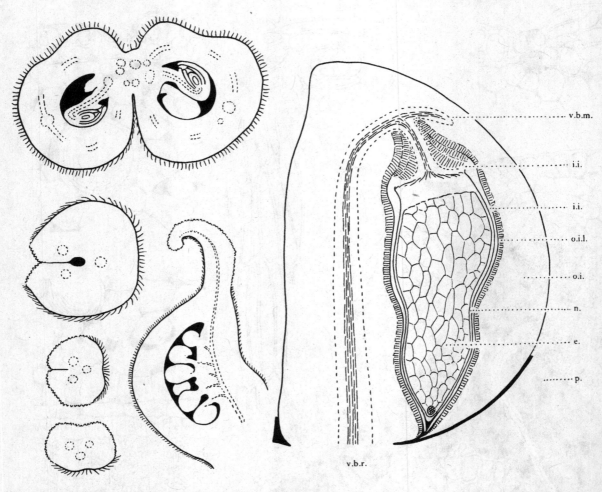

356　*Michelia champaka*. Carpel at anthesis in l.s., × 15. Carpel in t.s. at different levels up to the stigma, × 30.

Immature seed in l.s., × 30; o.i.l. the lignified endotesta, v.b.m. the postchalazal v.b., v.b.r. v.b. of raphe.

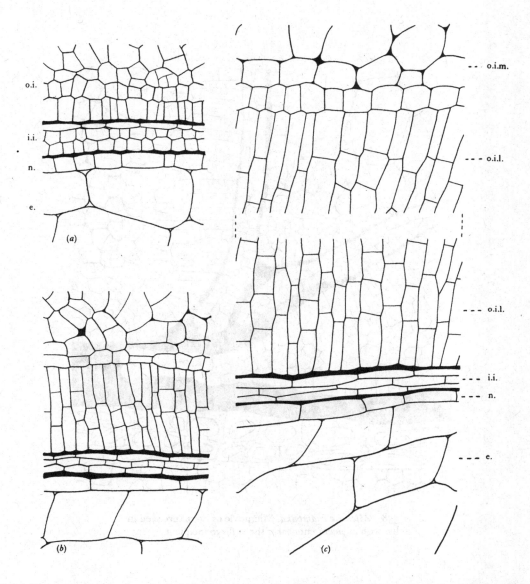

357 *Michelia champaka*: development of the integuments, ×225. (*a*) Ovule-wall in l.s. soon after fertilisation; (*b*) similar part in a half-grown seed; (*c*) similar part in a fully grown but immature seed, with fully multiplied endotesta. o.i.l. the lignified endotesta, o.i.m. mesophyll of o.i.

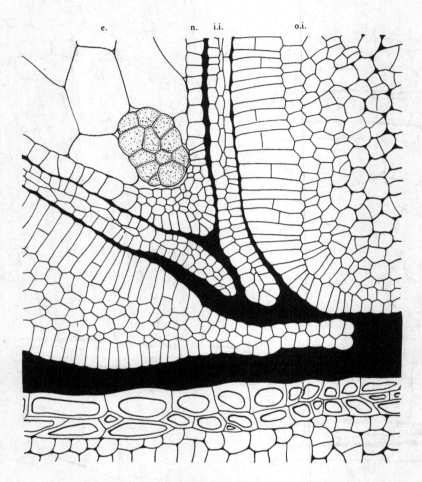

358 *Michelia champaka*. Micropyle of immature seed in
l.s. with adjacent endocarp, the embryo incipient, × 225.

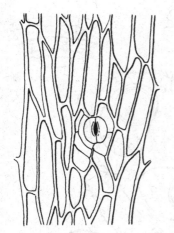
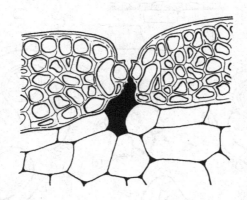
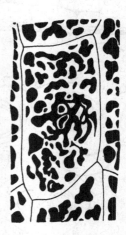

359 *Michelia champaka*. Stomata of the sarcotesta in surface-view, ×225; in section, ×400. Lignified cells of the endotesta with internal fibrils, the cell-spaces in black, ×800.

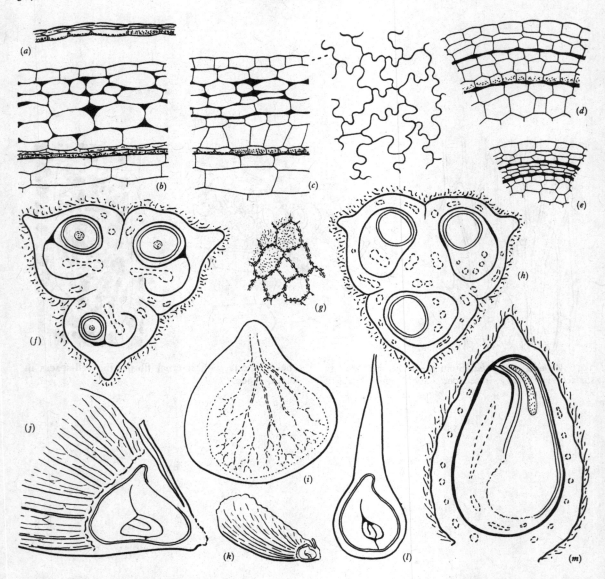

360 *Heteropterys angustifolia*. (*a*) Mature seed-coat in t.s., all the cell-layers collapsed on the endotegmen, ×225; (*b*) fully grown seed-coats in l.s. on the side of the seed with adjacent nucellus, the integuments indistinct and beginning to collapse on the endotegmen, ×225; (*c*) young seed-coat in l.s., the cells enlarged but the two integuments still distinct (testa 5 cells thick, tegmen 3 cells), with exotestal facets, ×225; (*d*) micropylar region of a half-grown seed in t.s. showing the distinct integuments (both 3 cells thick) and nucellus, with the endotegmen beginning to become pitted, ×225; (*e*) wall of ovule in t.s. ×225. (*f*), (*h*) Ovary in t.s. at the origin of the funicle and at the chalaza, ×25. (*g*) Endotegmic cells in surface-view, ×225. (*i*) Seed from the side of the brown chalaza, showing the ramifications of the raphe v.b. within the expanded chalaza, ×10; (*j*) ripe seed in l.s. in the samara with persistent stigma, ×5; (*k*) ripe samara in l.s., ×1; (*l*) the body of the ripe samara in l.s., ×5; (*m*) transmedian l.s. of a young seed in the carpel shortly after fertilisation, showing the extensive chalaza, ×25.

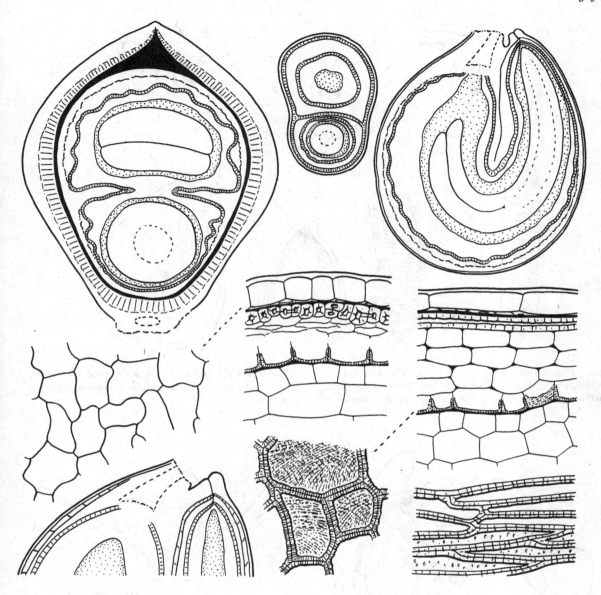

361 *Thryallis glauca*. Coccus of the fruit in t.s. with nearly mature seed (endocarp and endotegmen hatched), ×18. Immature seed in t.s. near the chalaza, showing the radicle and the septum of the tegmen (exotegmen and endotegmen hatched), ×18. Immature but fully grown seed in l.s. with the endotegmen (hatched) broken away, ×18. Micropyle and chalaza of the immature seed in l.s., ×25. Seed-coats in l.s. (right) and t.s., the inner cells of the testa crushed, with exotestal facets and exotegmic fibres, ×225; endotegmic facets, ×500.

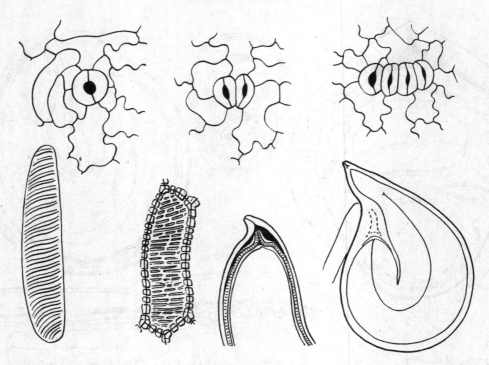

362 *Tristellateia australasiae*. Ripe seed in l.s., ×12. Micropylar end of the ripe seed, with the endotegmen hatched, ×25. Stomata of the testa, ×225. Endotegmic cells showing the outer surface with spiral thickening (left) and the inner scalariform facet (in optical section), ×500.

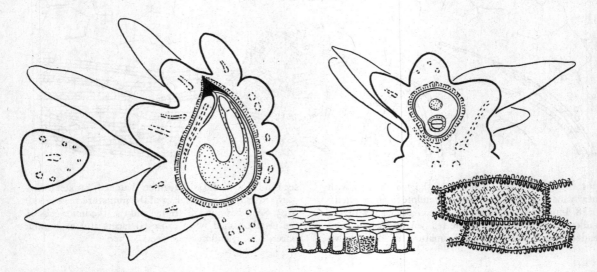

363 *Tristellateia australasiae*. Immature fruiting cocci in transmedian l.s. (left) and in t.s. (right), with the endocarp hatched, ×8; fruit-spine in t.s., ×10. Seed-coat in l.s., ×225. Cells of the endotegmen, ×500.

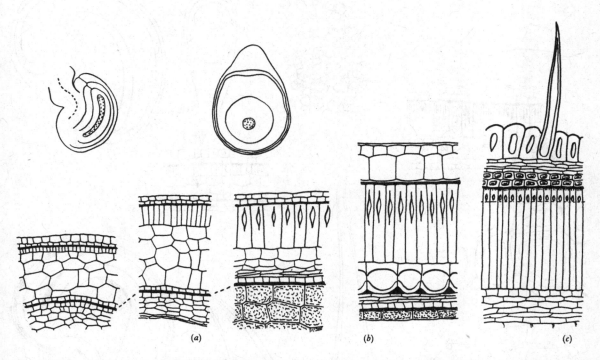

(a)

(b)

(c)

364 Malvaceae, seed-coats. (*a*) *Althaea officinalis*, ovule in l.s. and young seed in t.s. with enlarged nucellus, mag.; walls of ovule, immature seed, and mature seed, with endosperm replacing the nucellus, ×225. (*b*) *Lavatera* *trimestris*, seed-coats with endosperm in t.s., the cells of o.h. thick-walled and with wide air-spaces, ×225. (*c*) *Thespesia populnea*, seed-coats, ×130. ((*a*), (*b*)) After Guignard 1893; (*c*) after Reeves 1936.)

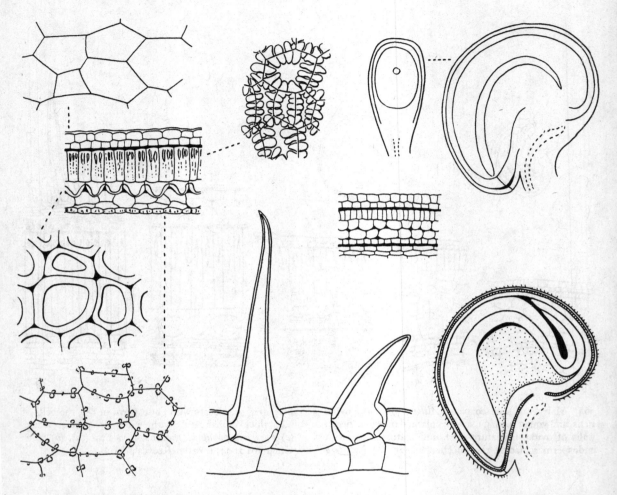

365   *Althaea rosea*. Ovule in median and transmedian l.s., × 25. Seed in median l.s., × 10. Wall of ovule and seed-coats in t.s., × 225; cell-facets and the epidermal hairs of the raphe, × 500.

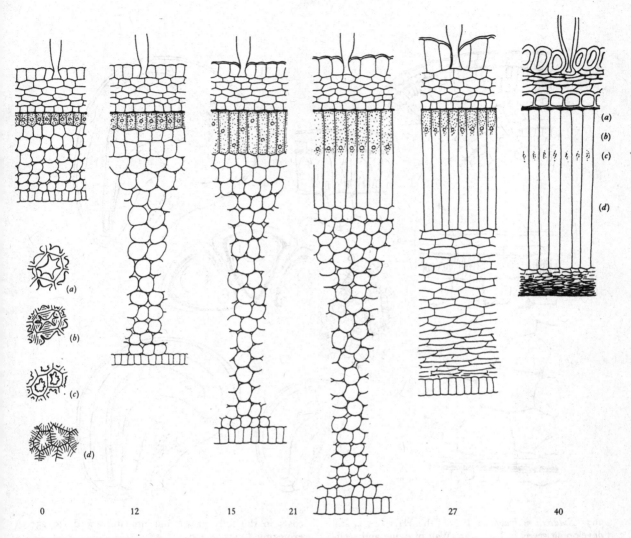

366 *Gossypium*, development of the seed-coats of Egyptian Cotton up to 40 days after flowering (from Balls 1915), × 300. (*a*)–(*d*), palisade-cells of the exotegmen at the levels indicated on the 40-day tegmen (after Reeves and Valle 1932), × 300.

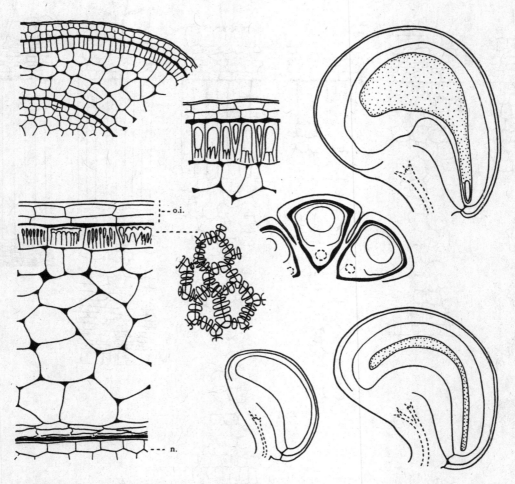

367 *Sidalcea malvaeflora*. Part of the ovary in t.s., and developing seeds in l.s., × 25. Wall of ovule and seed-coats in the fully grown but immature seed, × 225; exotegmic facets, × 400.

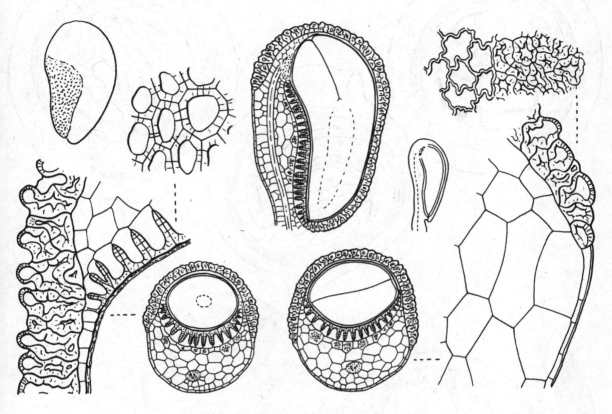

368   *Blakea trinervia.* Seed with dark thin-walled raphe, × 25. Seed in median l.s. and in t.s. across the hypocotyl and cotyledons, × 50; enlargements, × 225. Ovule, × 50.

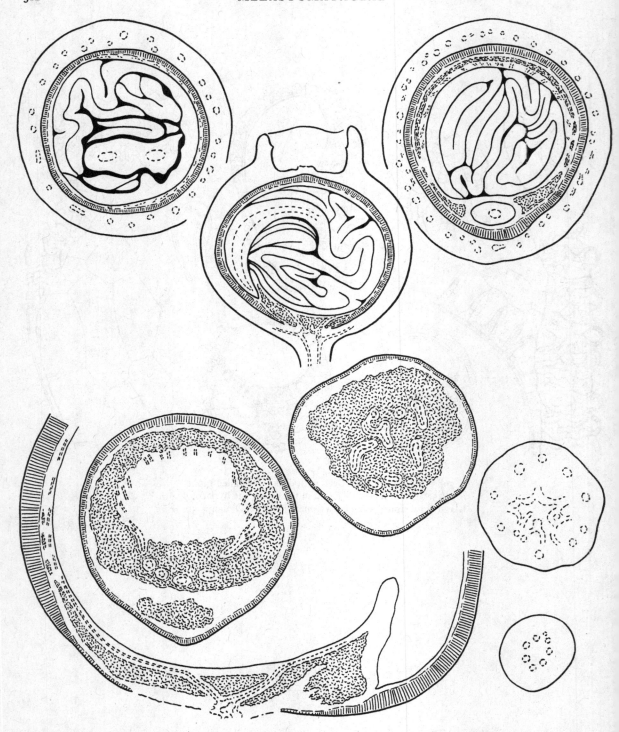

369   *Memecylon umbellatum*. Fruit in l.s., ×10; in t.s. of the upper half (left) and of the lower half (right), ×10. Hilar end of the seed in l.s. (×25) with t.s. at two levels (×18), and t.s. of the peduncle and base of the fruit (×12). Sclerotic tissue stippled.

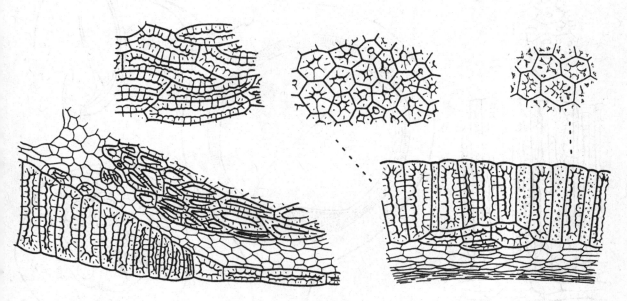

370 *Memecylon umbellatum*. Testa in t.s. with a group of sclerotic hypodermal cells, and in l.s. (lower left) at the micropylar base, × 125. Facets of the exotesta, × 225; upper left, from the longitudinally elongate cells round the hilum.

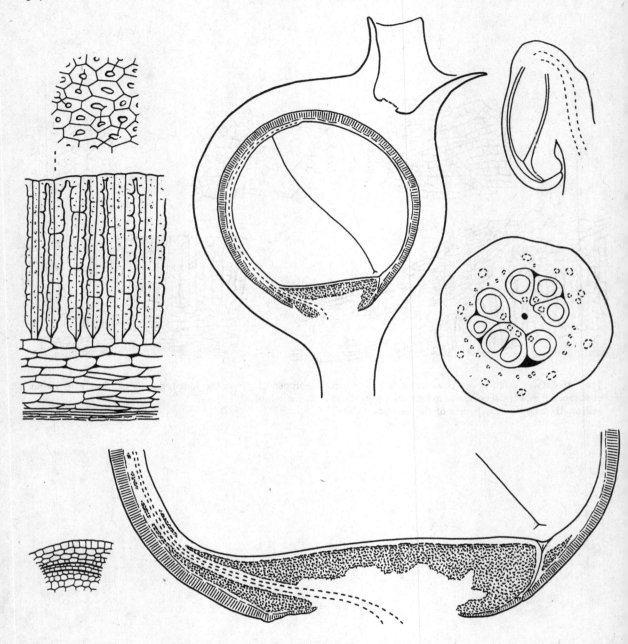

371 *Mouriria guianensis.* Fruit in l.s., × 8. Base of seed with micropyle in l.s., × 18. Ovary in t.s., × 25. Ovule in l.s., × 50. Wall of ovule and testa in t.s., × 225.

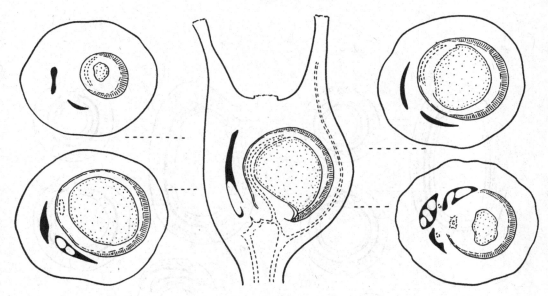

372 *Mouriria guianensis.* Immature fruits in l.s. and t.s., showing the development of the testa-palisade from the antiraphe-side of the growing seed, × 10.

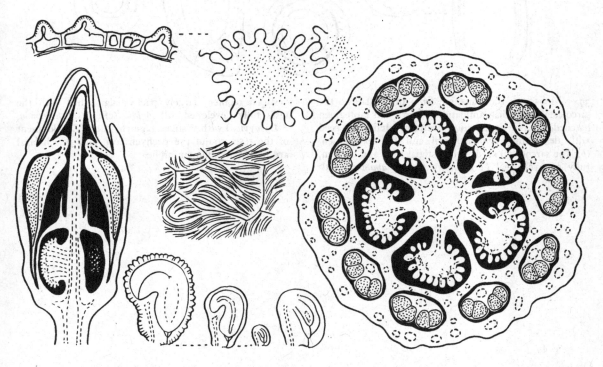

373 *Osbeckia octandra.* Flower-bud in l.s., × 10; base of flower-bud in t.s. with the anthers separated by receptacular septa, × 25. Ovule, young seed, and mature seed, × 50; ovule, × 125. Seed-coat (exotesta), × 225; exotestal facets, × 500. Cell of tegmen (?) with linear thickenings on the tangential wall, × 500.

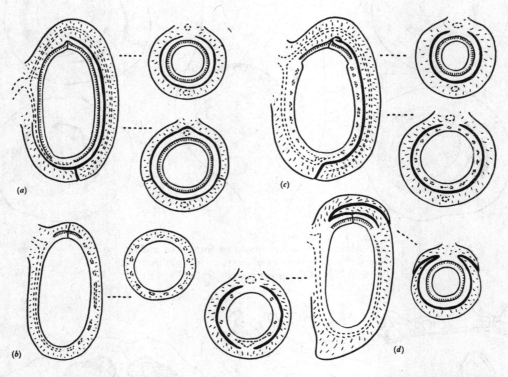

374 Meliaceae; the transition from the arillate to the sarcotestal seed; diagrammatic, with aril and exotegmen hatched. (a) The hypothetical primitive Meliaceous seed with descending funicular aril and full exotegmen. (b) The sarcotestal seed with vestigial testa and tegmen, fully pachychalazal (? *Guarea*). (c) *Aphanamixis* with vestigial tegmen, largely pachychalazal but with the testa fully developed. (d) *Aglaia* (cf. *A. trichostemon*, ? *Dysoxylon*) with vestigial tegmen, lateral development of the testa round the pachychalaza, and perichalazal aril from the raphe-antiraphe.

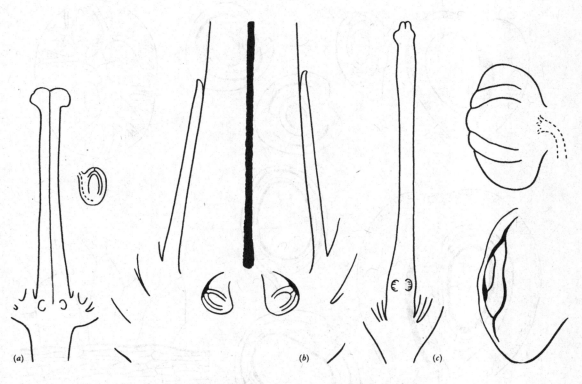

375 Meliaceae, ovaries. (a) *Dysoxylon cauliflorum*, ×6;
ovule, ×25. (b) *Sandoricum koetjape*, ×30. (c) *Chisocheton spicatus*, ×7; orthotropous ovule, ×75.

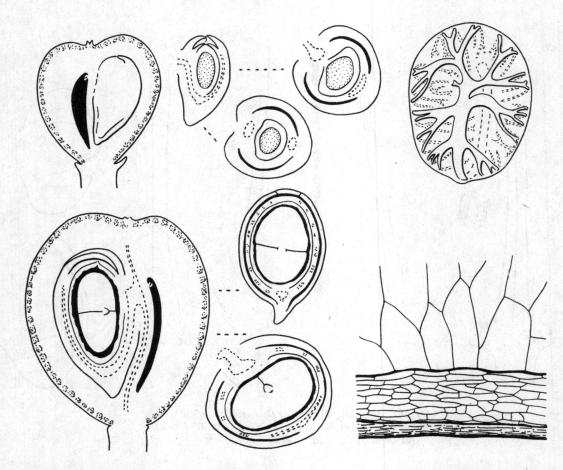

376 *Aglaia trichostemon*. Young and mature fruit in l.s., with sclerotic tissue in exocarp, × 5. Young seed in l.s., × 5; in t.s. at and below the hilum, showing the perichalazal structure, × 15. Mature seed in t.s. (× 6) and transmedian l.s. (× 5). Vasculature of the seed from the antiraphe-side, × 8. Seed-coats and inner tissue of the aril, the tegmen collapsed, × 75.

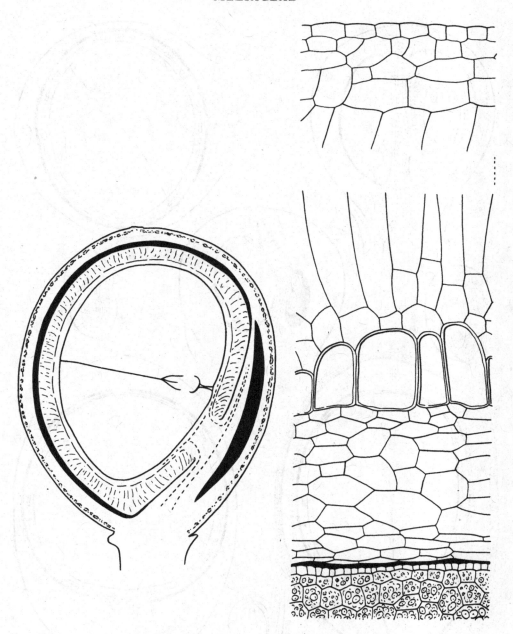

377 *Aglaia sp.* (Ceylon). Fruit (2-locular) in l.s., with
sclerotic tissue in exocarp, × 5. Seed-coat with sarcotesta
and cotyledon-tissue, × 225.

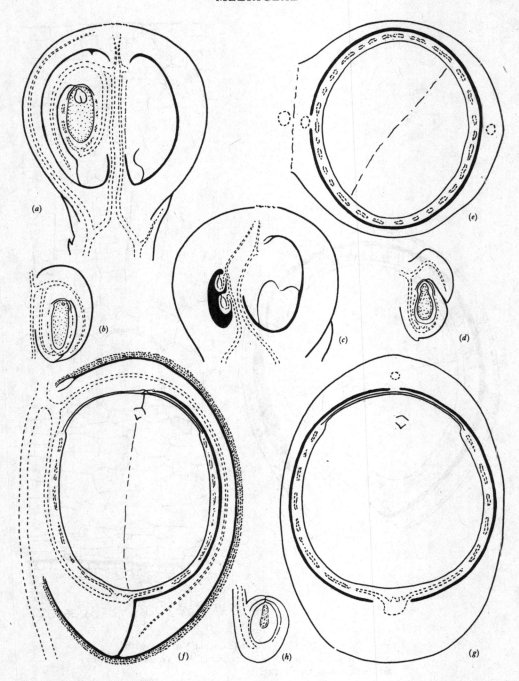

378 *Aphanamixis grandifolia*. (*a*) Young fruit in l.s. with one immature seed in l.s., showing the aril-outgrowths and vascular supply, × 5. (*b*) Younger seed in l.s., × 5. (*c*) Younger fruit in l.s., showing unfertilized ovules and aril-outgrowth on that developing into a seed, × 15. (*d*) The young seed in (*c*), in l.s., × 15. (*e*), (*f*), (*g*) Fully grown seed in t.s., median l.s., and transmedian l.s., showing the massive pachychalaza with v.b., the small apical part of the seed with free integuments and (*f*) the lignified endocarp (with the fibres cut transversely), × 4. (*h*) Ovule of open flower in l.s. showing the massive raphe-chalaza, × 50.

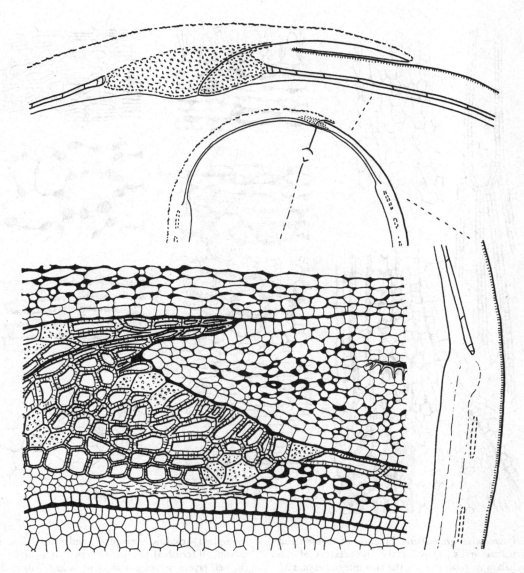

379 *Aphanamixis grandifolia*. Apex of seed in l.s. with the aril removed, × 5; with enlargement of the micropyle with woody endostome and fibrous exotegmen, × 25, and of the junction of the pachychalaza with the free integuments, the pachychalaza with three layers of tissue, × 25. Part of the micropyle and endostome in l.s. with adjacent cotyledon-tissue, × 120.

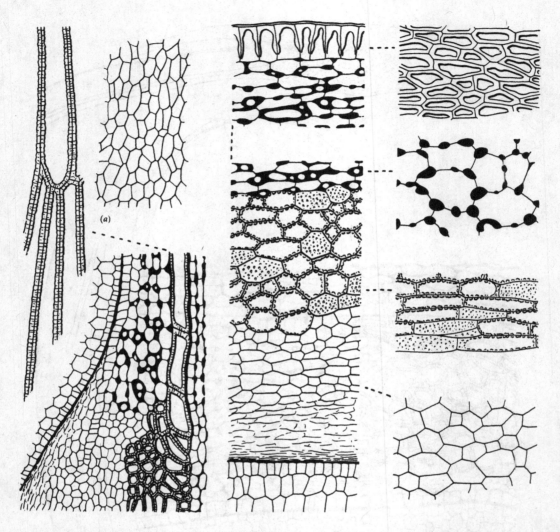

380 *Aphanamixis grandifolia*. Junction of tegmen and pachychalaza in l.s., showing the three layers of the pachychalaza in relation with the two integuments, and the adjacent cotyledon-tissue, × 120. Pachychalaza in t.s., with epidermal palisade, outer aerenchymatous mesophyll, sclerotic mesophyll, and thin-walled inner mesophyll (crushed towards the cotyledon-surface), and the cell-layers in tangential view, × 225. Fibres of the exotegmen in surface-view, × 225. (*a*) Surface of aril, × 120.

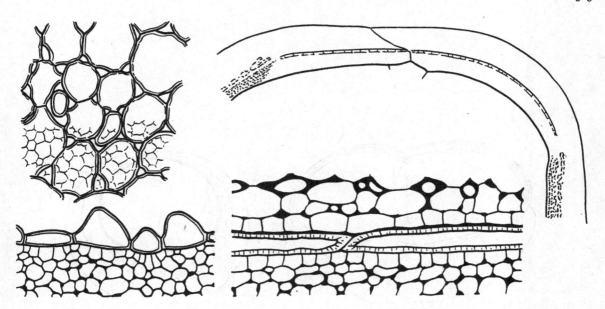

381 *Carapa guyanensis*. Micropylar area with adjacent embryo, showing the small free integuments adjoining the pachychalaza (hypostasial tissue stippled), × 5. Part of the integuments in section, showing the fibres of the exotegmen with the mesotegmen on the lower side, × 225. Epidermis of the pachychalaza in l.s. and in surface-view, × 120.

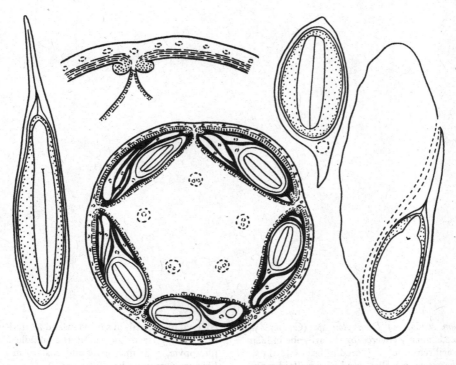

382 *Cedrela toona*. Seed in median l.s., × 5; in trans-median l.s. and t.s., × 10. Fruit in t.s., × 5; with enlargement of part of the pericarp at the junction of adjacent cocci, × 10.

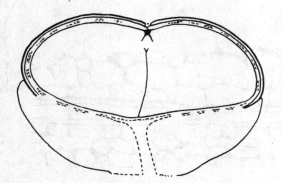
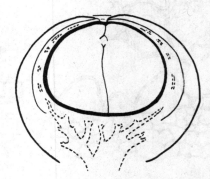

383 *Chisocheton divergens* (left) with arilloid base and
short aril and *C. sandoricocarpus* (right) exarillate; seeds
in l.s., ×4.

384 *Dysoxylon* seeds. (*a*) *Dysoxylon sp.* (Corner, 14
Nov. 1937), seeds mostly covered by the aril, one in hilar
view with the aril removed, ×2; seed in l.s., ×5; in t.s.,
×2. (*b*) *D. arborescens*, exarillate seed with arilloid raphe
in l.s. and t.s., ×3. (*c*) *D. cauliflorum*, seeds partly
covered by the aril, in side-view and hilar view, ×2; seed
in l.s., ×3; seeds in t.s. at the middle and near the
micropyle, ×8; micropyle and chalaza in l.s., ×8; the
fibrous exotegmen hatched.   ▷

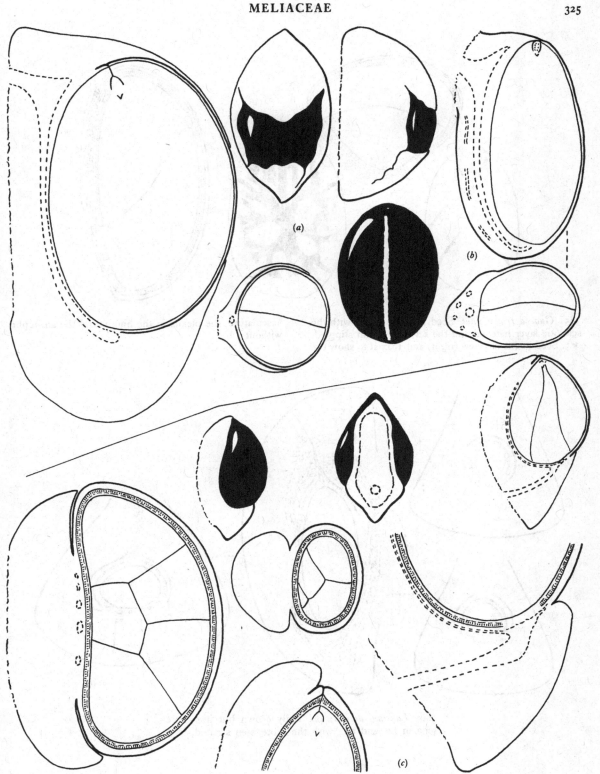

(a)

(b)

(c)

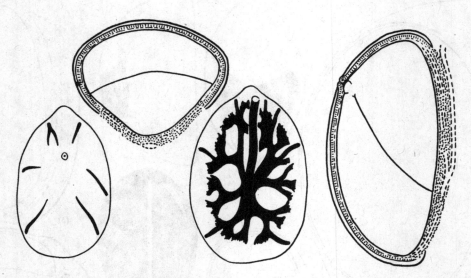

385 *Guarea trichilioides*. Seed in t.s. and l.s., with the sclerotic layer hatched and the sclerotic hilum stippled, × 5. Seed in adaxial view (right) and abaxial to show the vasculature, the plexus in the hilum, and the antiraphe without v.b.

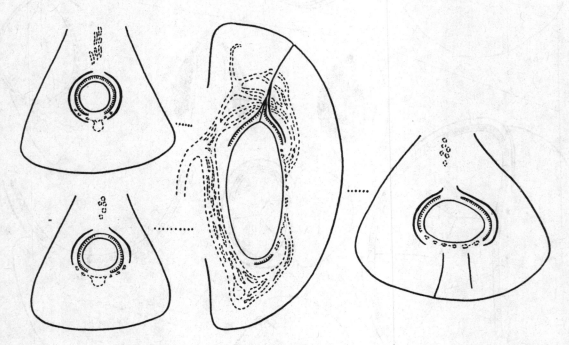

386 *Lansium domesticum*. Fully grown but immature seed in l.s. and t.s., with the exotegmen striated, × 3.

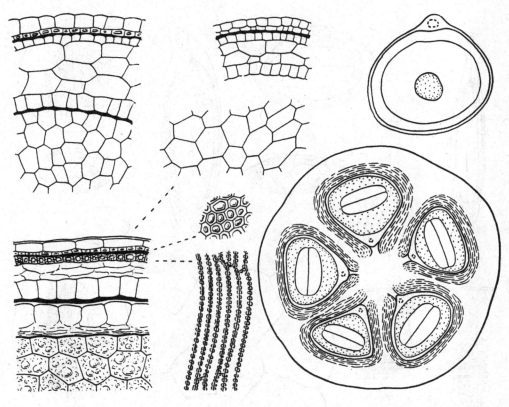

387 *Melia azedarach.* Fruit in t.s., × 10. Seed, immature but nearly full-sized, to show the large nucellus, × 25. Wall of ovule, immature and mature seed (with remains of nucellus), × 225.

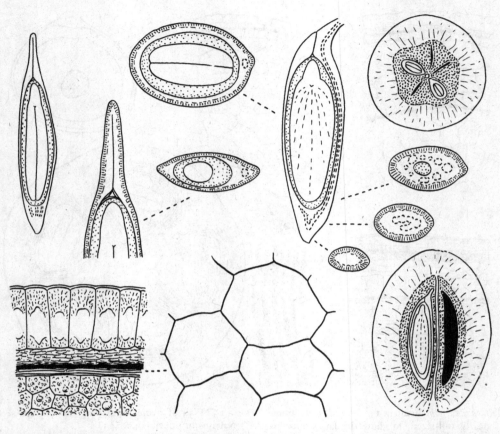

388 *Melia dubia*. Fruit in section with woody endocarp, ×2. Seeds in l.s. (×5) and in sectional detail (×10). Seed-coat structure in t.s. with endosperm, and tegmic cells in surface-view, ×225.

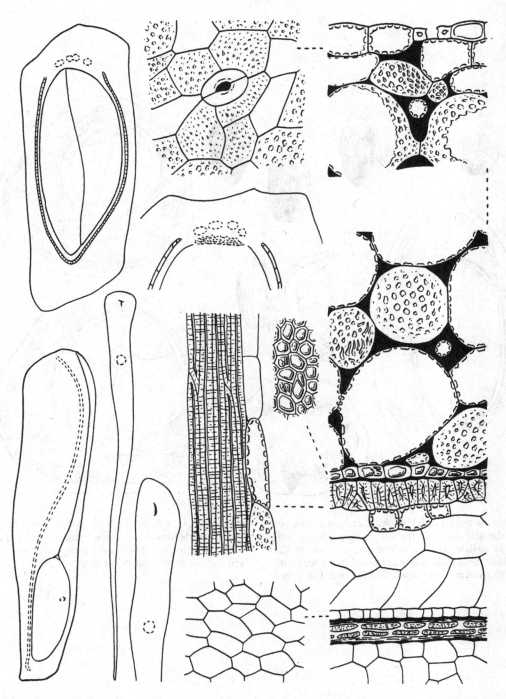

389 *Swietenia macrophylla*. Seed in l.s., × 1; in t.s., × 5; perichalaza, × 10. Seed-coat structure with 2 residual layers of endosperm and the cotyledon-surface, × 225.

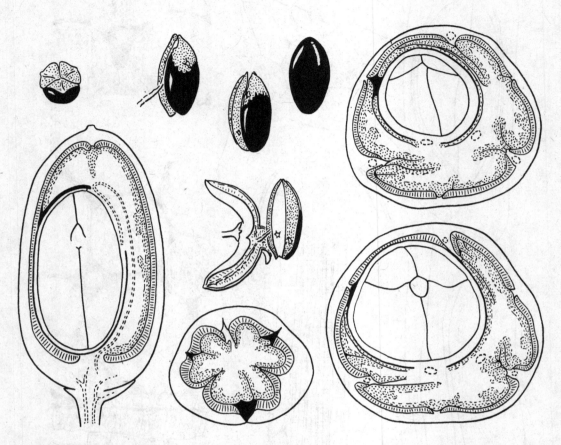

390 *Trichilia* sp. (Bolivia). Capsule and seeds, the red part of the aril stippled, the undeveloped ovules embedded in stellate recesses in the aril, the v.b. of the columella of the capsule acting as a funicle; one seed with the aril removed to show the white raphe, × 2. Unopened capsule in l.s., × 6; in t.s., × 10, at three levels from that of the radicle upwards, the cocci striated, the red cortex of the aril stippled, showing the compound nature of the aril and the loculicidal dehiscence.

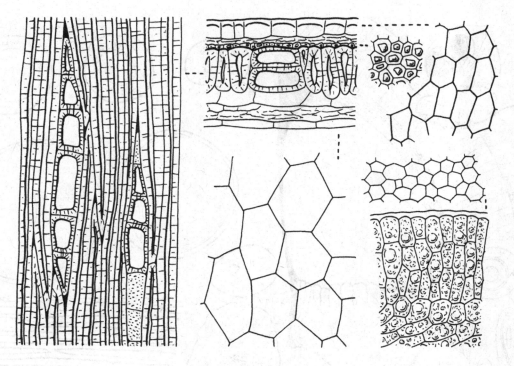

391 *Trichilia sp.* (Bolivia). Seed-coats in t.s., with facets of exotesta, endotesta, exotegmen and endotegmen; cortex of the aril (lower right); × 225.

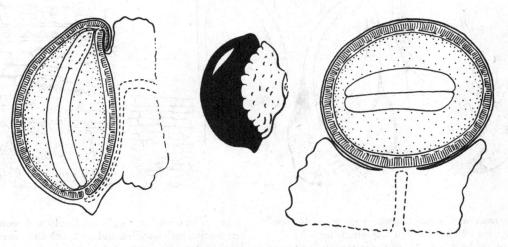

392 *Turraea pubescens*. Seed with aril in side-view, × 5; in l.s., × 8; in t.s., × 12; the exotegmen hatched.

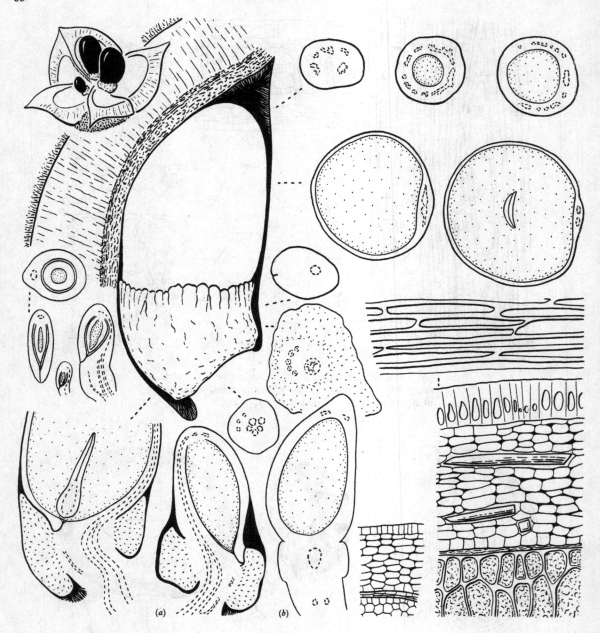

393 *Bersama abyssinica*. Capsule with 3 seeds, one abortive, × 1. Immature but full-sized seed in the loculus of the fruit, with l.s. of the micropylar end and t.s. at various levels from the chalaza to the funicle, × 5. Ovule and very young seed in l.s., × 5. Ovule in t.s., × 25; in transmedian l.s., × 10. Ovule-wall soon after fertilisation with nucellus, and testa with endosperm and crushed tegmen, × 225. (*a*) Unfertilized seed in l.s., × 5; (*b*) half-grown seed with incipient aril in transmedian l.s., × 5.

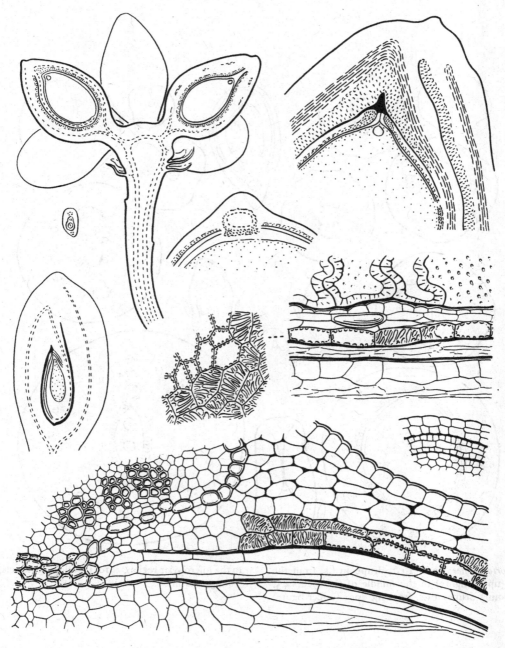

394 *Hortonia floribunda*. Fruit-cluster in l.s., × 2. Carpel of the open flower in t.s., × 10. Young drupe in l.s., × 10. Micropylar end of drupe in l.s., with endocarp and endotesta stippled, × 10. Perichalaza of the seed in t.s., × 25. Wall of ovule and seed-coats with adjacent endocarp, × 225. Part of the perichalaza of the immature seed with hypostase, integuments, and nucellus, × 225.

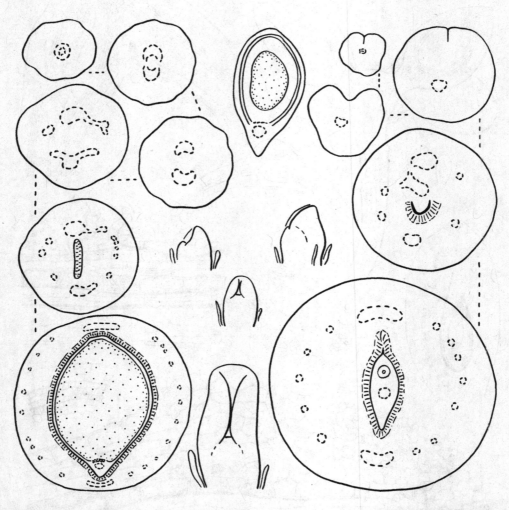

395 *Hortonia floribunda*. Left, drupe in t.s. from the
pedicel (upper figure) to the middle of the seed, × 5.
Right, young drupe in t.s. from the apex (upper figure)
to the micropylar region, × 10. Developing carpels, × 50.
Young seed in t.s., × 10.

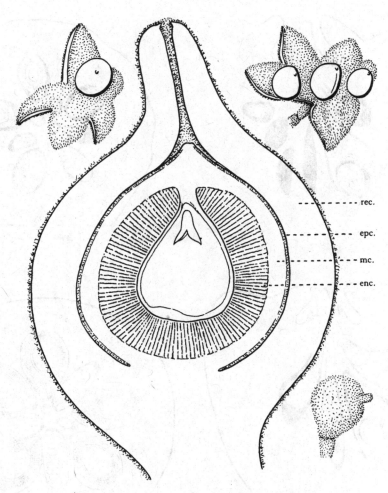

396 *Palmeria sp.* (NGF 12901). Dehisced fruits and
indehisced, × 1. Mature fruit with one seed in l.s., × 8;
epc. epicarp, mc. mesocarp, enc. endocarp, rec. receptacle.

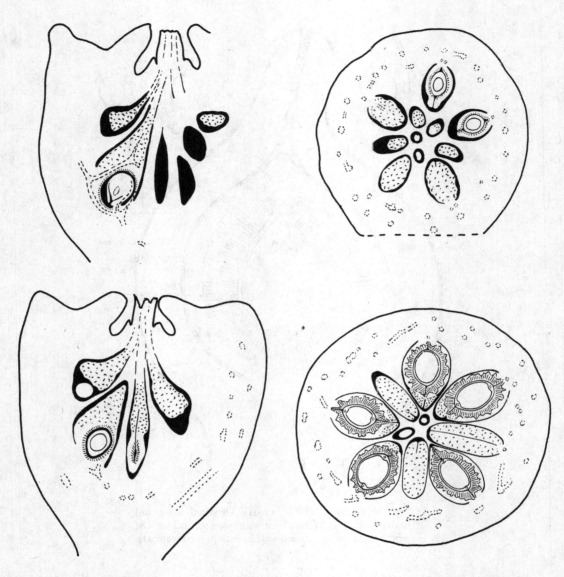

397 *Siparuna sp.* (Rio de Janeiro). Young fruiting receptacles (with style-remains) in l.s. and t.s., × 17.

Mature fruiting receptacle (lower right) in t.s., × 5. Carpellary aril speckled, endocarp striated.

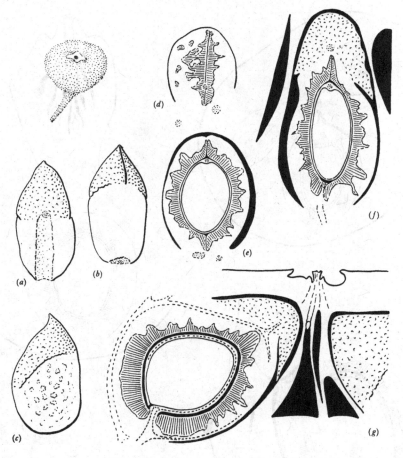

398 *Siparuna sp.* (Rio de Janeiro). Mature fruiting receptacle, × 2. (*a*)–(*c*) Ripe drupes: (*a*) showing hilum (attachment to receptacle); (*b*) showing the groove of the carpellary aril; (*c*) side-view, × 5. (*d*)–(*g*) Sections of drupes: (*d*) t.s. at the base of the endocarp; (*e*) t.s. near the base of the drupe; (*f*) transmedian l.s.; (*g*) median l.s., × 10. Carpellary aril speckled, endocarp striated.

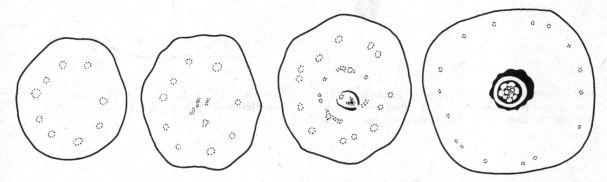

399 *Siparuna sp.* (Rio de Janeiro). Sections of the young receptacle from the peduncle (left) to the apex (right), showing the carpellary v.b. descending from the upper part of the receptacle, × 17.

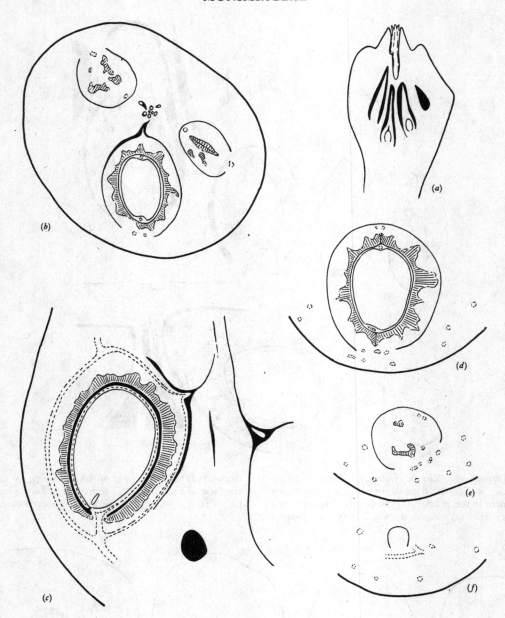

400 *Siparuna sp.* (Manaus). (*a*) Receptacle in l.s. shortly after pollination, × 17. (*b*) Fruiting receptacle in t.s. near the apex, × 10. (*c*) Ripe drupe enclosed in the receptacle, in l.s., × 10. (*d*), (*e*), (*f*) Drupe in t.s. at successive levels to the base (*f*), × 10. Endocarp striated.

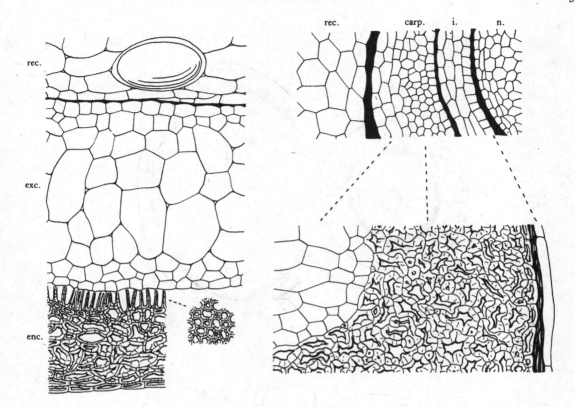

401 *Siparuna*, structure of drupe-walls, ×225. Left, *Siparuna sp.* (Mato Grosso) and right, *Siparuna sp.* (Rio de Janeiro), with the carpel at anthesis and maturity; rec. receptacular tissue, carp. carpellary tissue, i. integument, n. nucellus, exc. pulpy exocarp, enc. lignified endocarp.

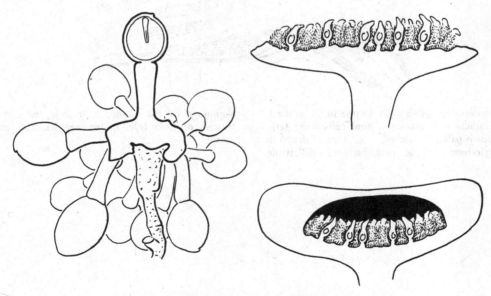

402 *Steganthera sp.* (RSS 151). Fruit-cluster in l.s., ×1. Female flower, open and in bud, in l.s., ×3.

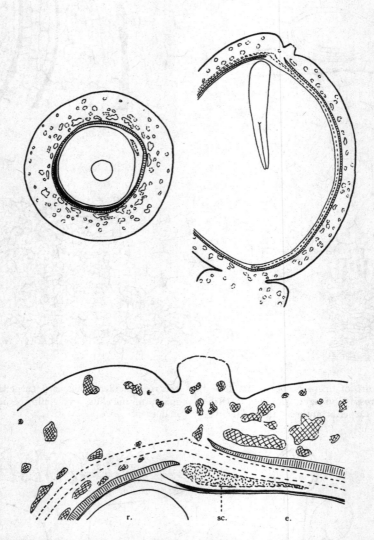

403 *Steganthera sp.* (RSS 151). Drupe in l.s. and t.s. across the radicle, with clusters of stone-cells in the pericarp, endocarp-palisade hatched, × 5. Apex of drupe in l.s. with style-base, × 25; sc. mass of sclerotic cells at the beginning of the raphe, r. radicle, e. endosperm; clusters of stone-cells in pericarp hatched; endocarp striated.

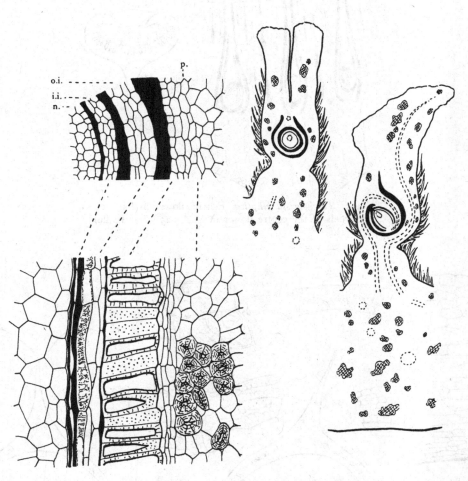

404 *Steganthera sp.* (RSS 151). Carpel of open flower in median and transmedian l.s., × 25; clusters of stone-cells in the carpel-wall and receptacle hatched. Wall of ovule and adjacent part of carpel, and seed-coats with adjacent endosperm and pericarp, × 225.

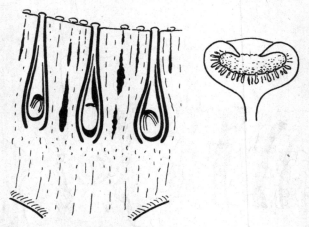

405 *Tambourissa elliptica.* Female flower in l.s., ×2. Carpels of the centre flower in l.s., ×25 (after Baillon 1871).

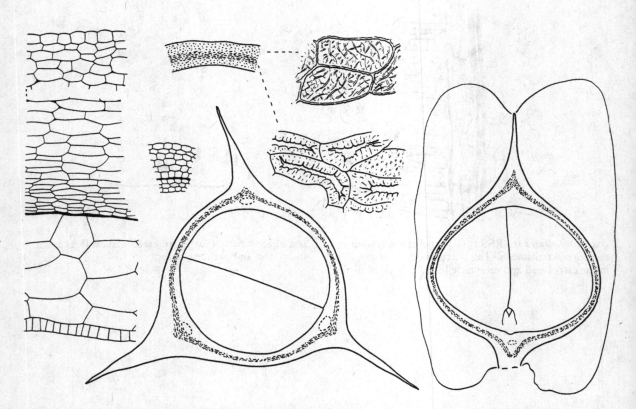

406 *Moringa oleifera.* Ripe seed in l.s. and t.s., with sclerotic mesotesta, ×5. Wall of ovule and the seed-coats of an immature seed (testa shown in part), ×225. Dia-gram of the testa of a mature seed in t.s., ×10, with cell-details, ×225.

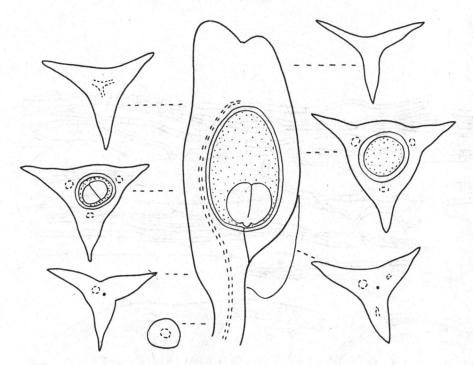

407   *Moringa oleifera*. Young seed in l.s. and t.s. at
various levels, ×5.

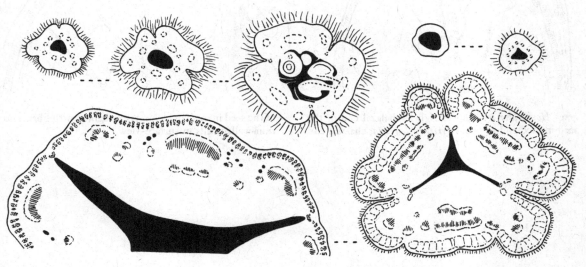

408   *Moringa oleifera*. Ovary in t.s. of the upper part, stigma and style (upper right) in t.s., ×25. Nearly mature fruit in t.s. between the seeds (lower left), with sclerotic exocarp, ×5. Young fruit in t.s. between the seeds (lower right), with incipient sclerotic tissue, ×10.

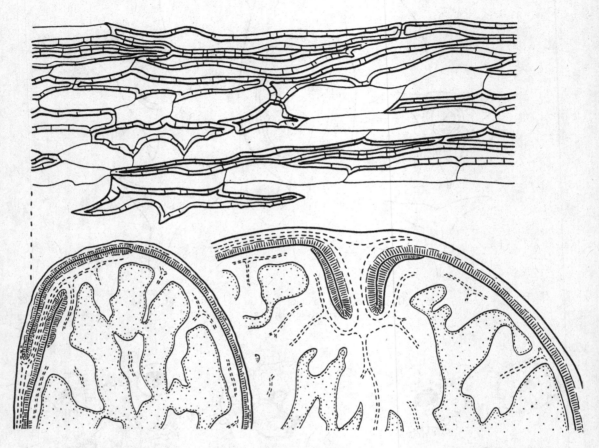

409 *Horsfieldia irya* (lower right); chalazal end of the seed in median l.s., × 8. *Myristica lowiana*, chalazal end of the seed in median l.s., × 3; exotegmen with fibres and thin-walled cells in surface-view, × 225.

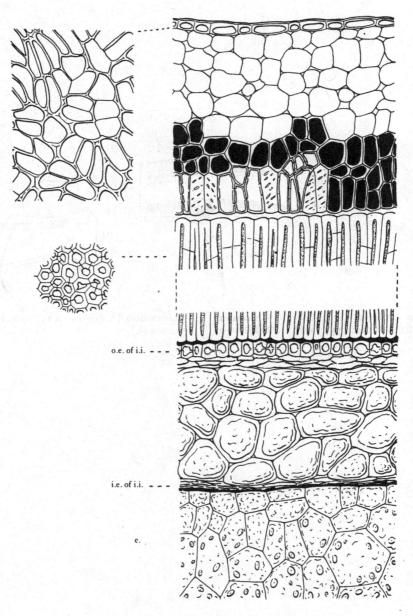

o.e. of i.i.

i.e. of i.i.

e.

410 *Horsfieldia subglobosa* var. *brachiata*. Mature seed-coats in t.s., with tannin-cells (black) in the inner part of the testa; exotegmic fibres cut transversely; × 225.

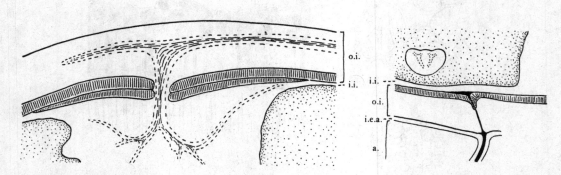

411 *Horsfieldia subglobosa* var. *brachiata*. Chalaza in median l.s. (left) and micropyle with arillostome in transmedian l.s. (right), ×15; i.e.a. inner epidermis of the aril (a.).

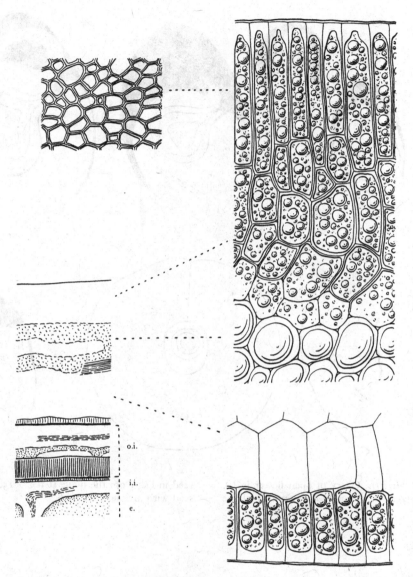

412 *Horsfieldia subglobosa* var. *brachiata*. Aril (right) in t.s., with cell-details, ×225. Diagram of the aril and seed-coats (left) in t.s. to show the relation of the parts of the aril, ×30.

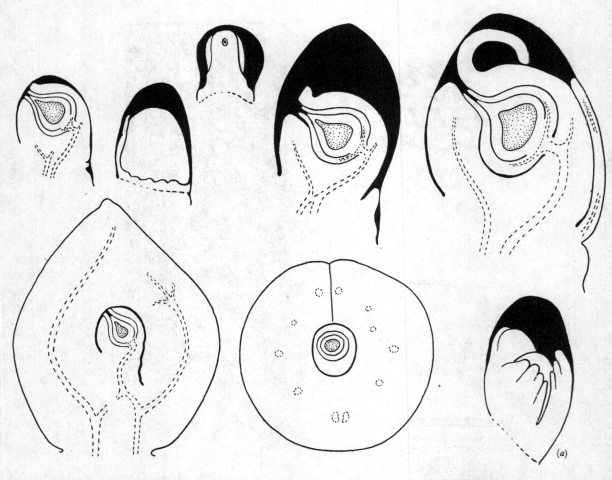

413 *Horsfieldia sylvestris.* Ovary of open flower in l.s. and t.s., × 15. Stages in the early development of the seed in l.s., from the ovule (left), × 15. (*a*) Very young seed with incipient aril, × 6.

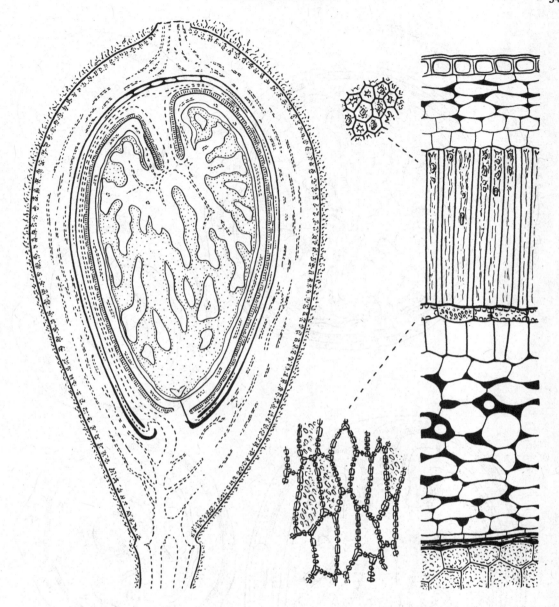

414 *Knema laurina*. Fully grown but immature pericarp and seed (with trace of nucellus) in median l.s., × 8. Testa, tegmen and outer part of the endosperm in t.s., with facets of the exotestal palisade and the exotegmic sclerotic cells; the nucellus crushed; × 225.

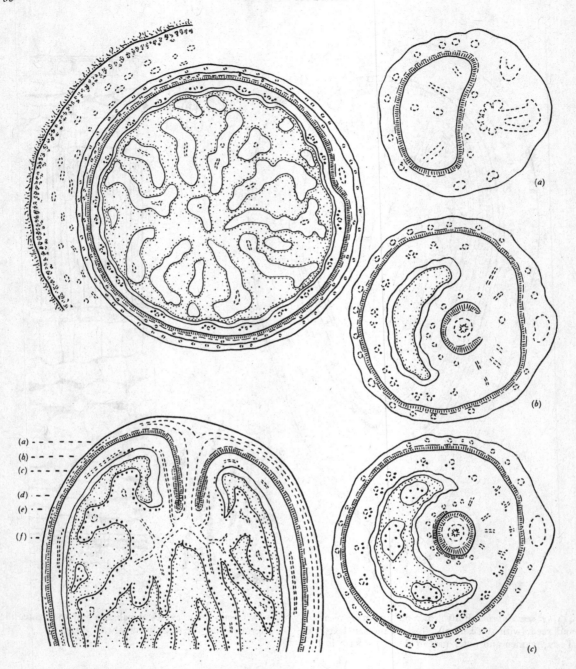

415 *Knema laurina.* Fully grown but immature seed with a trace of the nucellus in t.s., with part of the pericarp (sclerotic cells in exocarp), showing the vascular aril, testa and tegmen with ruminations, × 8. Chalazal end of the seed in median l.s., showing the levels of the transverse sections (a)–(c) and (Fig. 416) (d)–(f), with the oil-cells in black, × 8. Chalazal end of the seed in t.s. at the levels (a)–(c), with oil-cells in black, × 12.

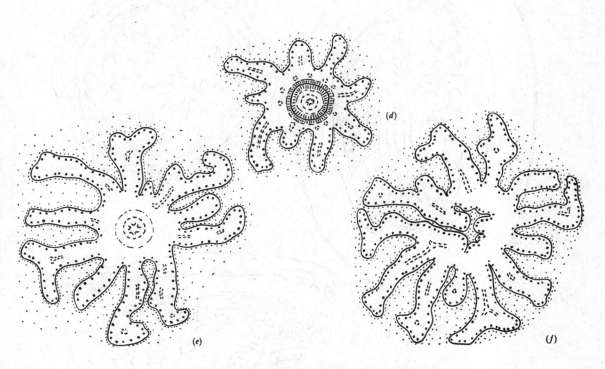

416   *Knema laurina.* Chalazal tube in t.s. (*d*) and the
ruminations at the levels (*d*)–(*f*) (Fig. 415), in t.s., with
oil–cells in black, × 12.

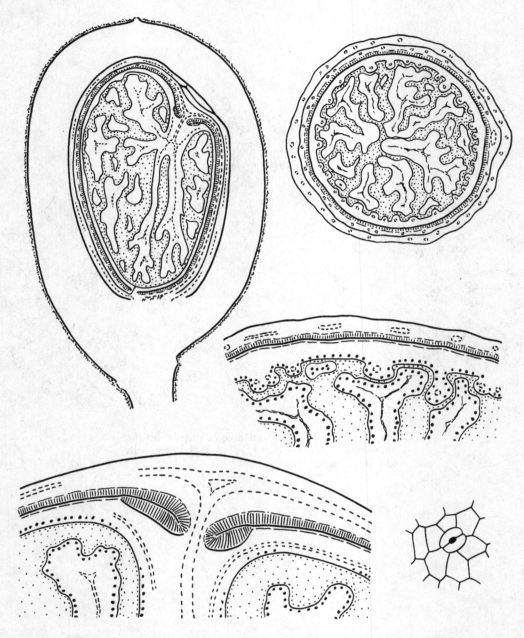

417   *Knema sp.* (Java). Fully grown but immature fruit in median l.s., × 3. Seed in t.s. with vascular aril, testa and tegmen, the free ruminations with collapsed central tissue, the tegmen with longitudinal ridges, × 4. Chalazal end of the seed in median l.s. and part of the seed-coat with outer part of endosperm in t.s., the oil-cells in black, × 12. Stoma of the exotesta, × 225.

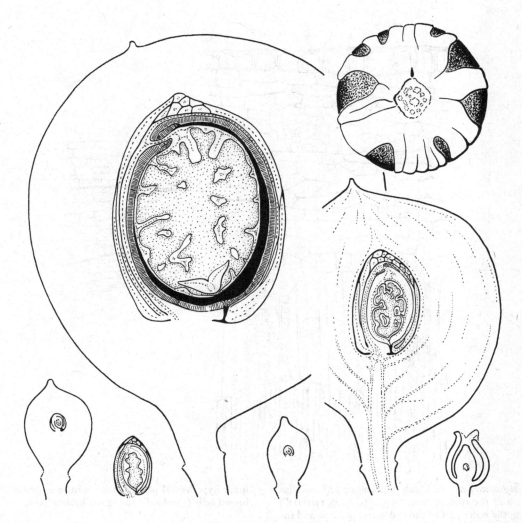

418 *Myristica fragrans.* Fruit-development in l.s. from the flower (lower right), with the seed in hilar view with arillostome, × 2. Endotestal palisade striated, just forming at the chalaza in the penultimate stage.

419 *Myristica fragrans.* Ovule-wall (right) with nucellus in t.s., and the mature testa in t.s. (Ceylon material), showing the outer part of the endotestal palisade and the inner hypodermal palisade with groups or trabeculae of brown cells (speckled) among the hyaline cells, ×225.

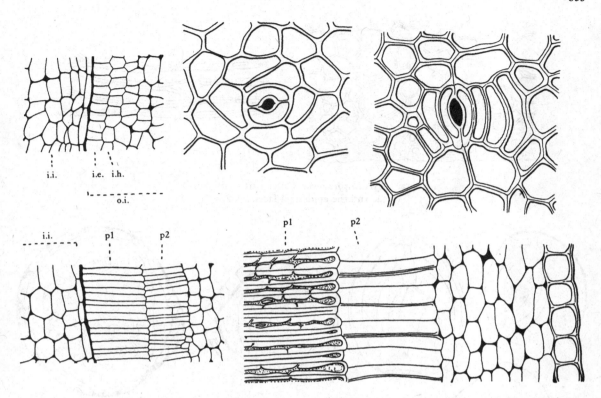

420 *Myristica fragrans.* Junction of testa and tegmen in young seeds (left) to show the development of the two palisades (p1, p2) from i.e. and i.h. of the testa, × 225.

Testa in t.s. (lower right) with the outer part of the endotestal palisade (Singapore material), × 225. Stomata of the exotesta, × 400.

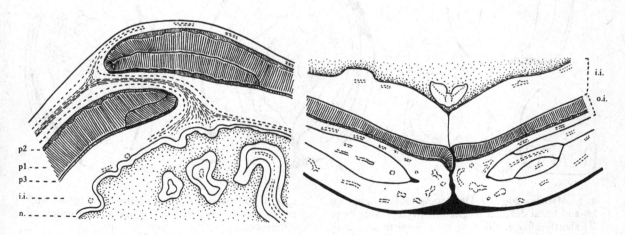

421 *Myristica fragrans.* Chalaza (left) in median l.s., and the micropyle with arillostome (right) in transmedian l.s., from fully grown but immature seeds, × 7.

Endotestal palisade (p1), inner hypodermal palisade (p2) and fibrous exotegmen (p3) striated.

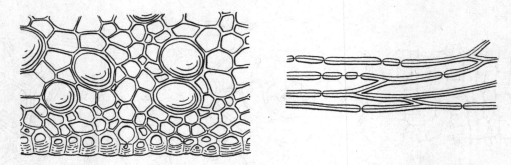

422   *Myristica fragrans.* Outer part of the aril in
t.s. and the epidermal facets, × 225.

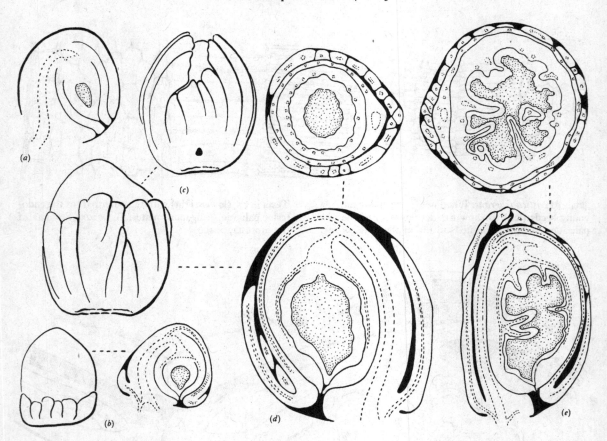

423   *Myristica fragrans.* Developing seeds in side-view,
l.s. and t.s. at different magnifications. (*a*) Ovule, × 30;
(*b*) shortly after fertilisation with incipient aril, × 15;
(*c*) surface-views of (*d*) showing arillostome and beginning of ruminations of the endosperm, × 15; (*e*) half-
grown seed covered by the aril but without endotestal
differentiation, × 7.

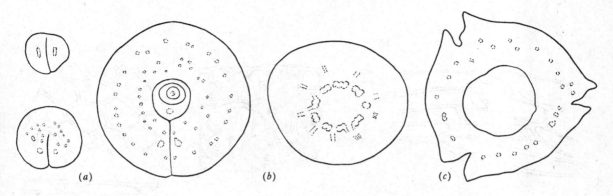

(a) (b) (c)

424 *Myristica fragrans.* Ovary in t.s. at the stigma and style (*a*), at the ovular region, and at the base (*b*), with the base of the perianth (*c*), ×15.

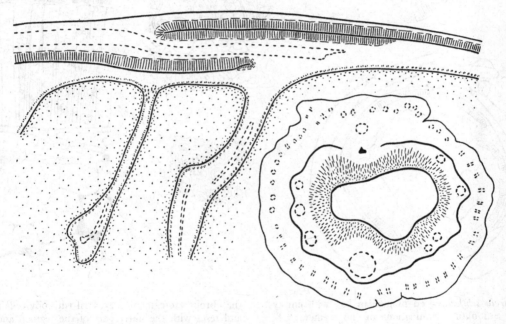

425 *Myristica iners.* Chalaza in l.s., showing the counter-palisade of the tegmen, ×10. Seed in t.s. at the base just above the separation of the funicular part of the aril, with the endotestal palisade cut obliquely, ×5.

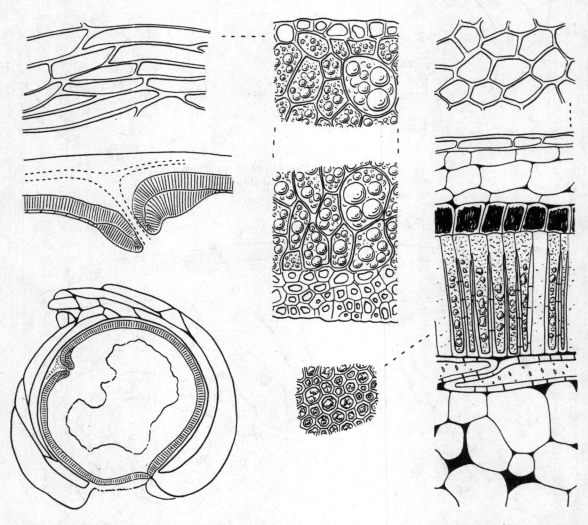

426 *Virola sebifera*. Seed in median l.s., with empty interior and obtuse ruminations of the tegmen, × 8. Chalaza in l.s. with the endotestal palisade leading into the fibrous exotegmen, × 25. Aril with oily cells in t.s., and testa with the outer part of the tegmen and cell-details, × 225.

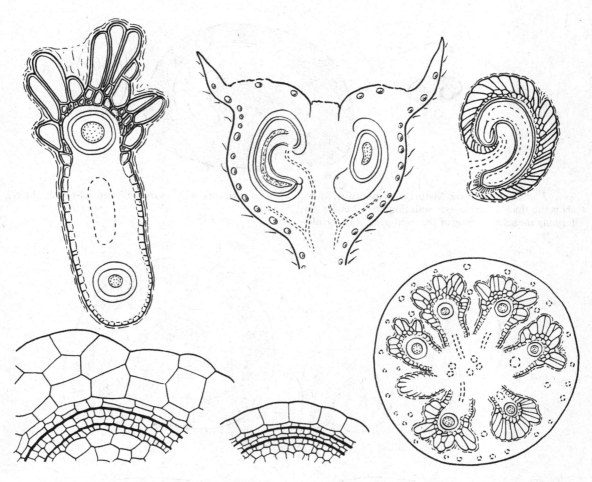

427 *Decaspermum fruticosum*. Ovary soon after fertilization in l.s., with hypodermal oil-glands, × 25. Ripe seed in l.s., enclosed in the fibrous endocarp, × 10. Immature fruit in t.s., × 10. Immature seed in transmedian l.s., enclosed in the thin fibrous endocarp, × 25. Wall of ovule and of young seed (with multiplicative exotesta), × 225.

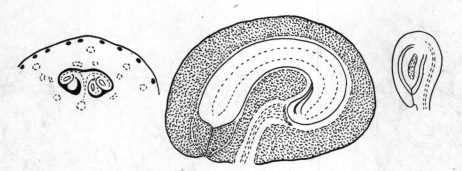

428 *Psidium cattleyanum.* Mature seed in median l.s. to show the thick sclerotic testa with thin-walled endotesta forming the sclerotic plug of the micropyle, × 10. Ovule, × 50. Placenta with ovary-wall and oil-glands (black), × 10.

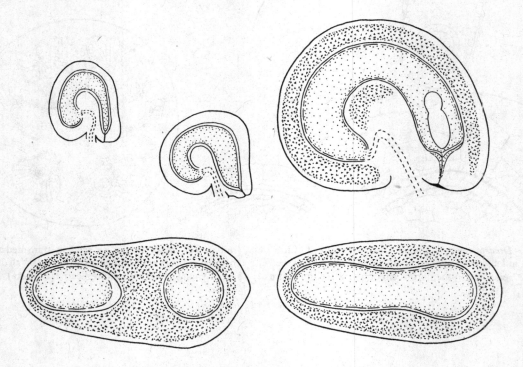

429 *Psidium guajava.* Young seeds, soon after fertilization, and fully grown but immature seeds, × 15; the sclerotic tissue of the mesotesta speckled.

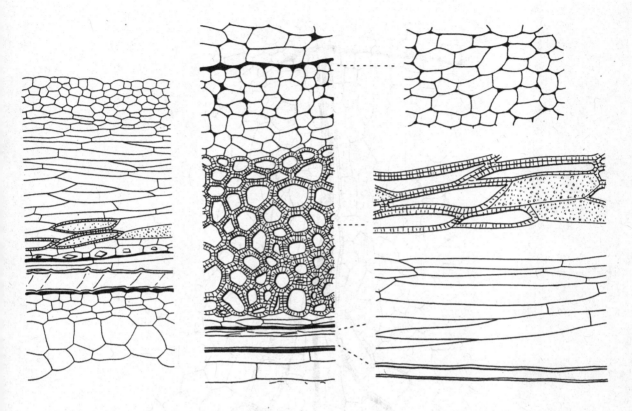

430 *Psidium guajava*. Left, seed-coats of a young seed in t.s. near the chalaza, and testa with incipient sclerotic tissue, the endotesta as crystal-cells, × 225. Right, the nearly mature seed-coat in l.s., with adjacent endocarp and nucellar remains, endotesta collapsed, × 225.

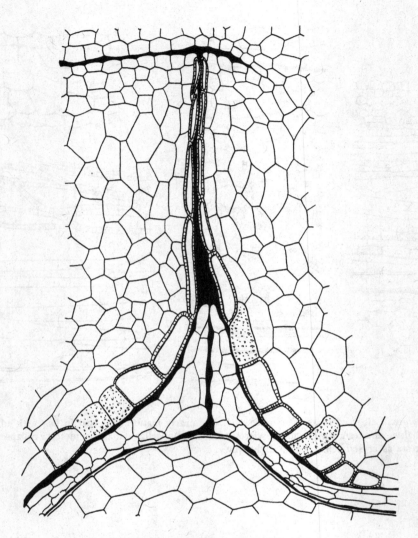

431 *Psidium guajava*. Micropyle of the fully grown but immature seed in median l.s., the endotesta sclerotic, with nucellar remains, × 225.

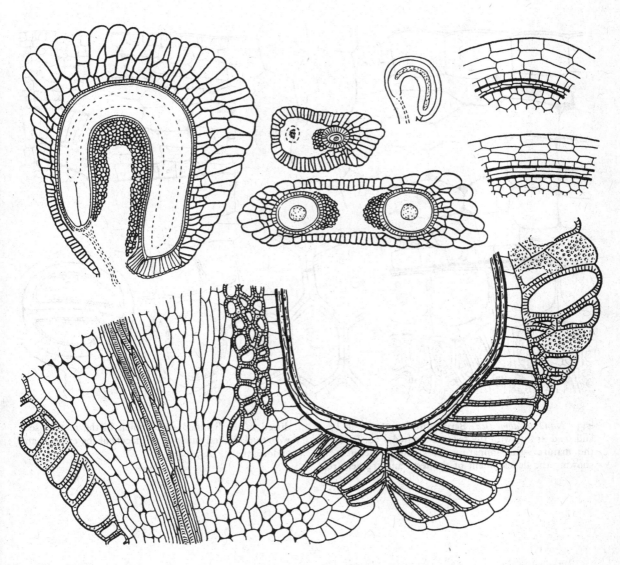

432 *Rhodomyrtus tomentosa*. Mature seed in median l.s., showing the sclerotic tissue of the testa and sinus, × 25. Immature seed in t.s. above the micropyle and across the middle part of the seed, × 25. Ovule, × 25. Ovular in-teguments and nucellus in t.s. (above) and l.s., × 225. Micropyle and hilum in median l.s., showing the endo-testal palisade of the micropyle with remains of tegmen and nucellus, × 120.

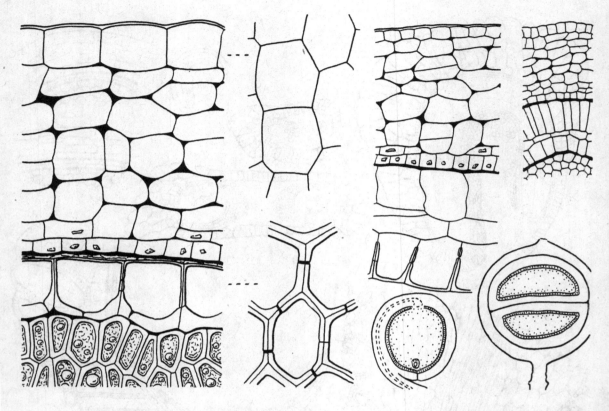

433 *Nandina domestica*. Berry with two seeds in l.s., and seed in t.s., with the tegmen striated, ×4. Parts of the mature seed, immature seed, and ovule in t.s., showing the development of the testa and tegmen (re- duced to the lignified endotegmen in the mature seed with thick-walled endosperm); facets of the exotesta and endotegmen, ×225.

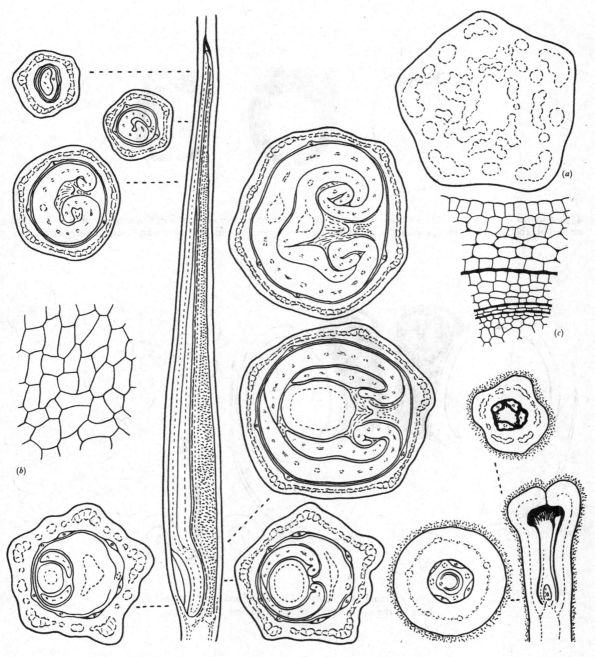

434 *Pisonia longirostris*. Flower-bud in l.s. (× 10) and t.s. across the ovule (× 25) and across the staminodes and stigma (× 10). Basal part of the fruit (anthocarp) in l.s., the perianth-tube with a continuous ring of fibro-vascular bundles, the brownish raphe-ridge stippled, × 2; sections across the fruit, × 5. Pedicel (*a*) just below the base of the ovary to show the complex vasculature, × 8. Ovary-wall, integuments and nucellus in t.s. (*c*), × 225. Outer epidermis of the testa (*b*) in surface-view, × 225.

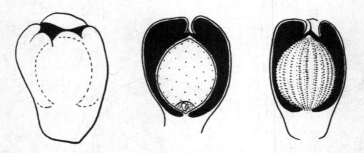

435 *Victoria amazonica.* Seeds with aril enclosing the air-bubble (black) in side-view, median l.s., and with the aril cut away on one side; the small embryo surrounded by small-celled endosperm and massive perisperm; × 10.

436 *Ochna kirkii.* Fruit in l.s., × 5. Carpels and style-base in the open flower, × 25. Style and ovary in t.s., × 25.

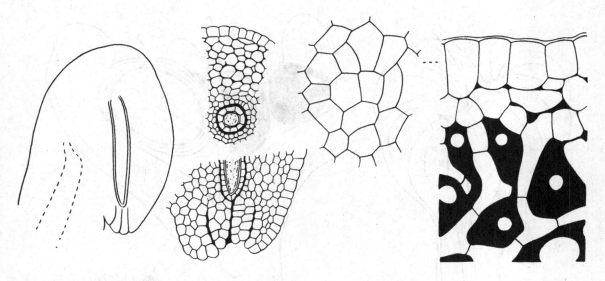

437 *Ochna kirkii*. Ovule at anthesis in l.s., ×110. Ovule-wall in t.s. and micropyle with vestigial integu- ments, ×225. Outer part of the mature testa in t.s., ×225.

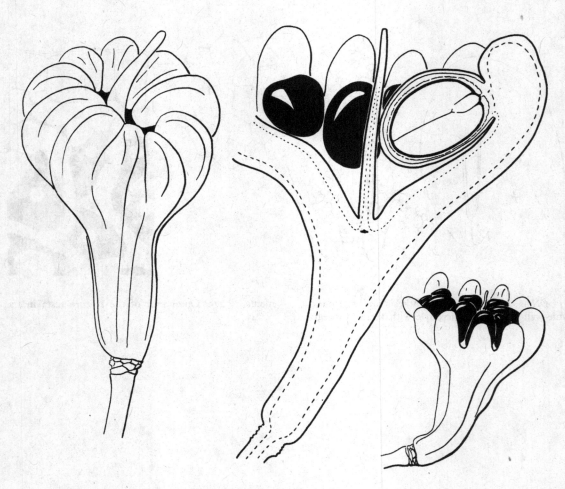

438 *Ouratea sp.* (Manaus, Corner 137). Young fruit,
×4. Mature fruit, ×2; in l.s., ×3.

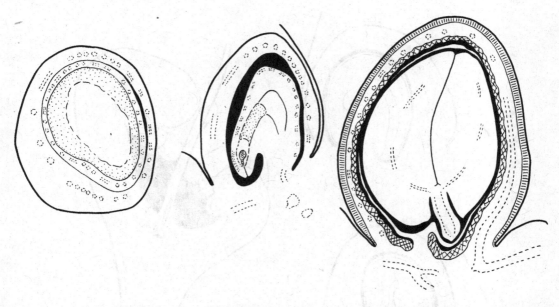

439 *Ouratea sp.* (Manaus, Corner 137). Young and mature drupes in l.s., the outer palisade of the drupe striated, the fibrous endocarp hatched; young drupe, ×25; mature drupe, ×8. Young drupe (left) in t.s. with the cellular endosperm stippled and with non-cellular endosperm, ×25.

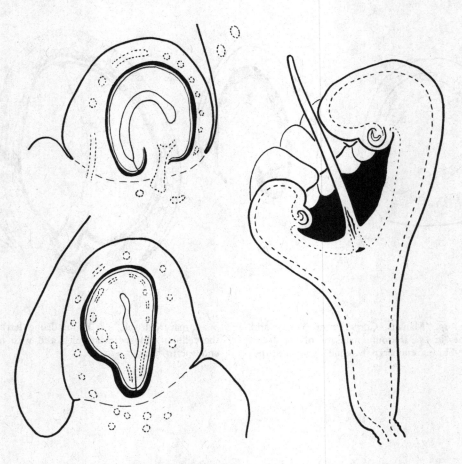

440 *Ouratea sp.* (Manaus, Corner 137). Young fruit in l.s., with marginal incurved seeds, × 4. Young drupes in median and transmedian l.s., × 25.

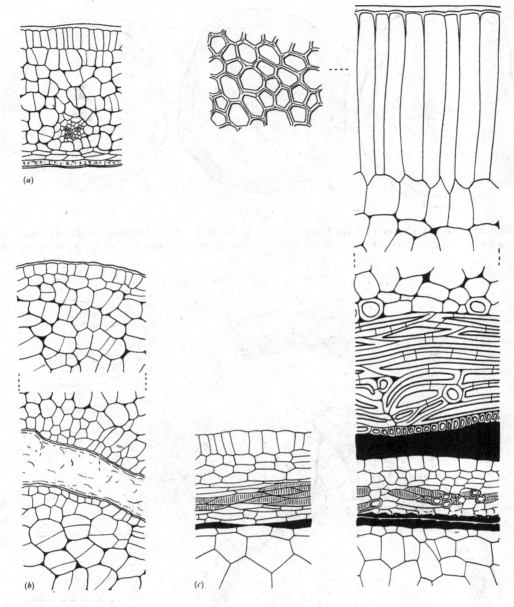

441 *Ouratea sp.* (Manaus, Corner 137). Pericarp, testa and endosperm of the mature drupe (right); (*a*) pericarp of the very young drupe; (*b*) ovule-wall in l.s. with endo- sperm (central) shortly after fertilization; (*c*) seed-coat with endosperm in l.s. of half-grown seed; × 225.

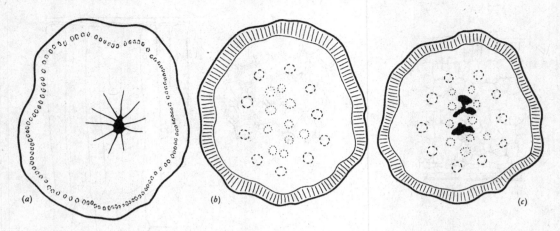

(a)   (b)   (c)

442   *Ouratea sp.* (Manaus, Corner 137). (a) Receptacle in t.s. at the level of the central stylar cavity, ×8; (b) style in t.s. with v.b. and fibrous bundles (? stylar canals, dotted), ×50; (c) base of style in t.s., as in (b), ×50.

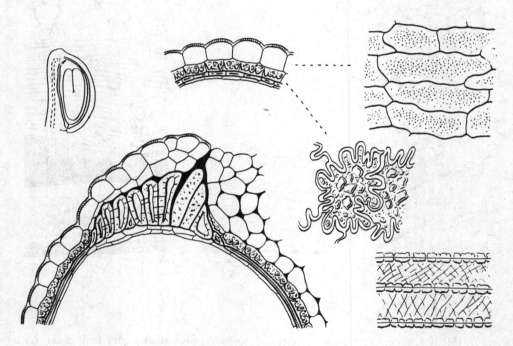

443   *Jussieua peruviana*. Seed in l.s., ×25. Seed-coats in t.s. with exotestal facets, ×225; endotestal facets and exotegmic fibres, ×500. Micropyle in l.s., ×225.

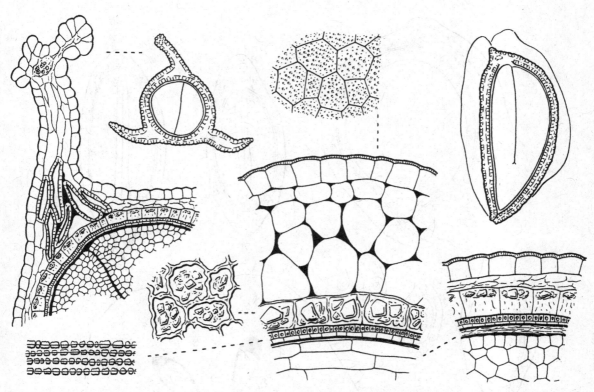

444 *Oenothera sp.* Seed in l.s. and t.s., showing exotesta and endotesta, × 25. Wing of seed in t.s., with sclerotic cells in the testal mesophyll, × 120. Seed-coats in t.s. of the immature seed (with nucellar tissue) and of the mature seed (with cotyledon-tissue), × 225; fibres of the exotegmen in surface-view, × 500.

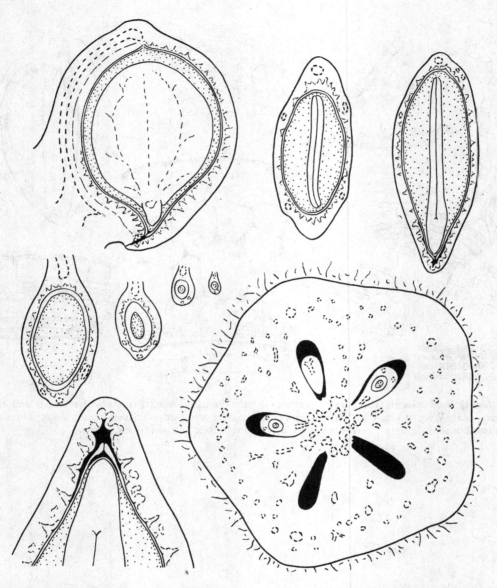

445 *Averrhoa bilimbi*. Sections of nearly mature seed, × 10. Ovule and young seed in t.s., × 10. Micropyle of the nearly mature seed in transmedian l.s., × 25. Ovary in t.s., × 25.

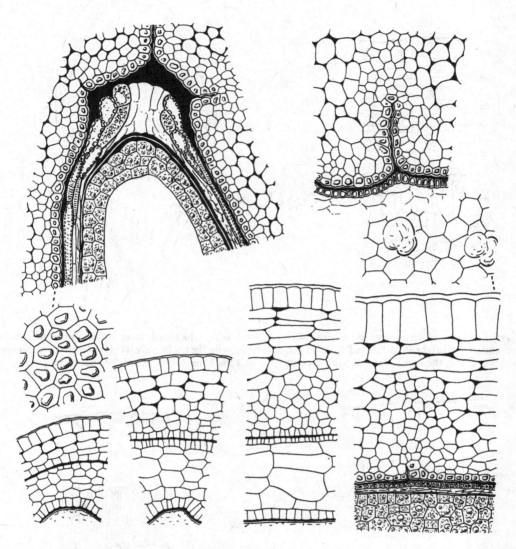

446 *Averrhoa bilimbi*. Endostome, surrounded by exo-stomal tissue, of the fully grown seed, and a 'rumination' of the testa near the micropyle, × 225. Wall of ovule and developing seed-coats in t.s., the oldest with endosperm, × 225. Crystal-cells of the endotesta in surface-view, × 500.

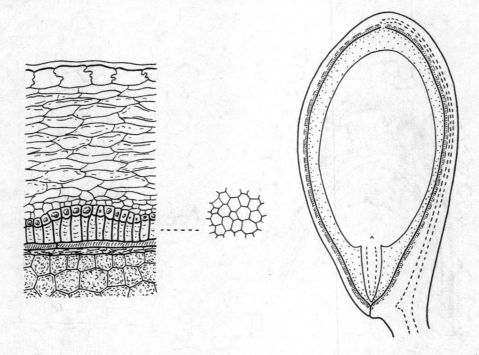

447 *Averrhoa carambola.* Mature seed in l.s. with endosperm and endotesta (hatched), × 8. Seed-coats in l.s., the mesotesta crushed, the endotestal cells with crystals and much thickened inner walls, the tegmen crushed except for the tracheidal exotegmen with spiral thickening, and the outer cell-layers of the endosperm, × 225.

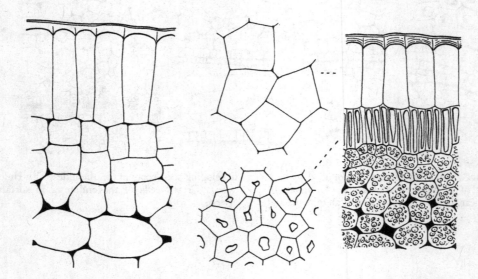

448 *Paeonia arietina* (right); outer part of mature testa in t.s., with aqueous exotestal palisade and lignified hypodermal palisade, the cells of the mesotesta thin-walled and full of starch, × 90; cell-facets, × 225. *P. delavayi* (left); outer part of the immature testa with the exotestal palisade and incipient hypodermal palisade, × 225.

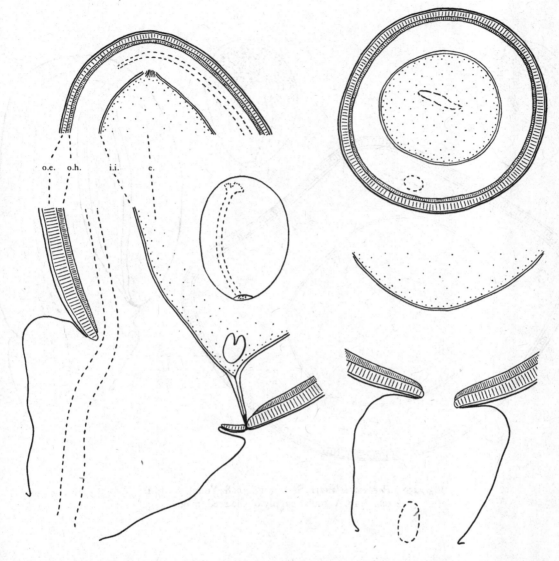

449 *Paeonia arietina*. Mature seed in t.s., chalaza in l.s., micropyle in l.s., and seed-base with funicle in trans-median l.s., the exotesta and o.h. (testa) striated, × 18. Vascular supply to the seed, mag.

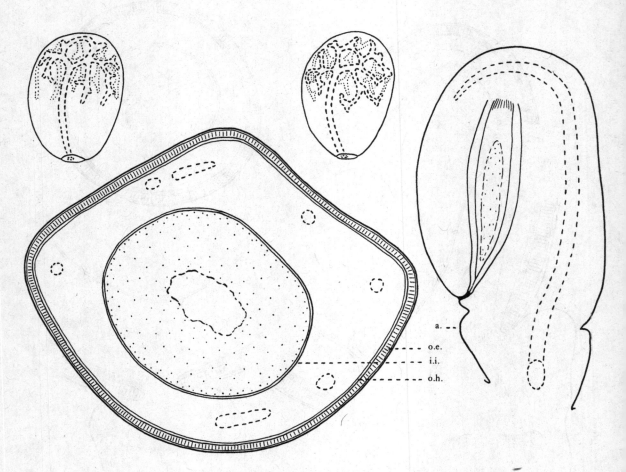

450 *Paeonia delavayi*. Seed in t.s., × 8. Young seed in
l.s., × 18. Vascular supply of the seed, mag.

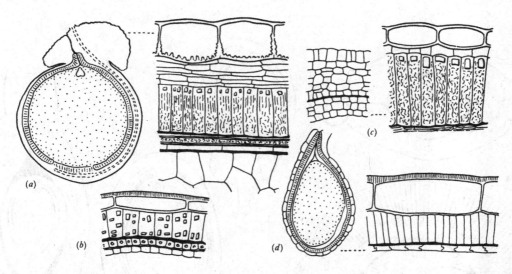

451 Papaveraceous seed-coats. (*a*) *Dendromecon rigida*, seed, × 10; seed-coats, × 140 (Berg 1966). (*b*) *Macleaya microcarpa*, mag. (Röder 1958). (*c*) *Sanguinaria sp.*, ovule-wall and seed-coats, × 200 (Shaw 1904). (*d*) *Argemone mexicana*, seed × 90; seed-coats, × 180 (Sachar 1955).

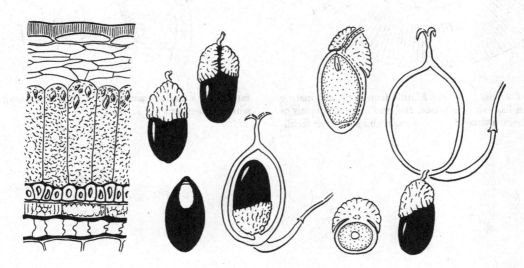

452 *Bocconia frutescens*. Fruits and seeds, × 3. Seed-coat in t.s., with the single effete layer of the nucellus, mag. (after Meunier).

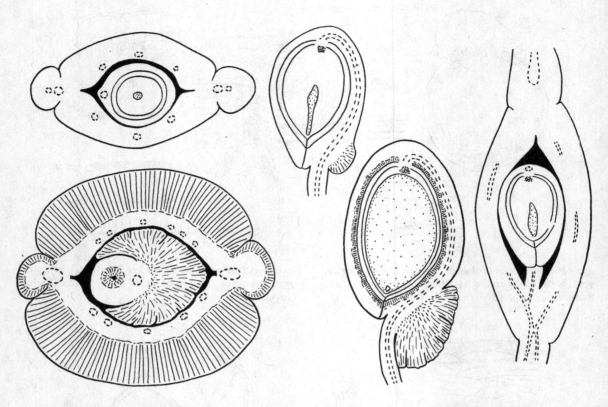

453 *Bocconia frutescens*. Mature fruit in t.s. and mature seed in l.s., with endotesta, aril, and outer epidermis of the pericarp striated, × 12. Young fruit soon after fertilization in t.s. and transmedian l.s., and the young seed with incipient aril in l.s., × 25.

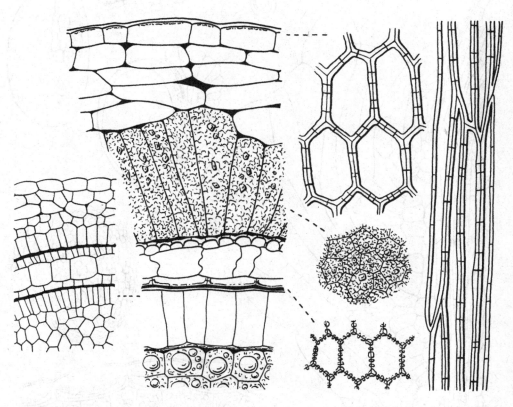

454   *Bocconia frutescens*. Ovule-wall in t.s. soon after
fertilization, and mature seed-coats in t.s. with endosperm, × 225. Facets of exotesta, endotesta, exotegmen
(fibres, on the right) and endotegmen, × 225.

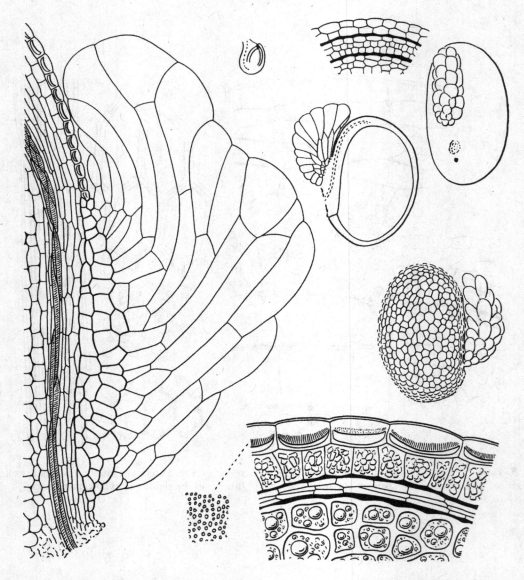

455 *Chelidonium majus.* Ovule at anthesis and mature seeds, ×25. Ovule-wall and seed-coats with endosperm in t.s., ×225; pores of the outer wall of the exotesta, ×400. Raphe with aril in l.s., ×225.

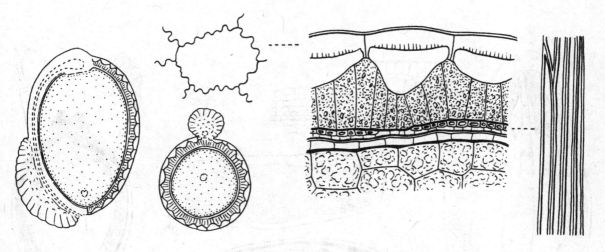

456   *Macleaya cordata*. Seed in l.s. and t.s., × 25. Seed-
coat with outer part of the endosperm in t.s., × 225.

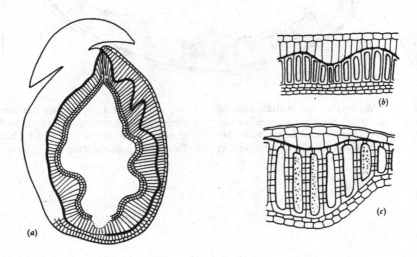

457   *Passiflora* (after Kratzer 1918). (*a*) *P. suberosa*,
young seed with aril in l.s. (*b*) *P. cuprea*, testa and tegmen.
(*c*) *Paropsia obscura*, testa and tegmen.

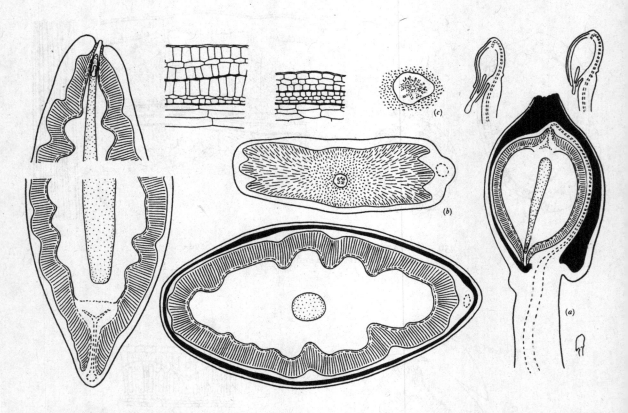

458 *Passiflora edulis.* Micropylar and chalazal ends of the fully grown but immature seed in transmedian l.s., ×25. Fully grown but immature seed in t.s. with surrounding aril, ×25. Wall of ovule and of young seed in l.s., ×225. (*a*) Fully grown but immature seed with funicle and aril, and ovule with protruding nucellus, ×8; sections of ovule, ×25. (*b*), (*c*) Chalaza in t.s., ×25. Exotegmic palisade striated.

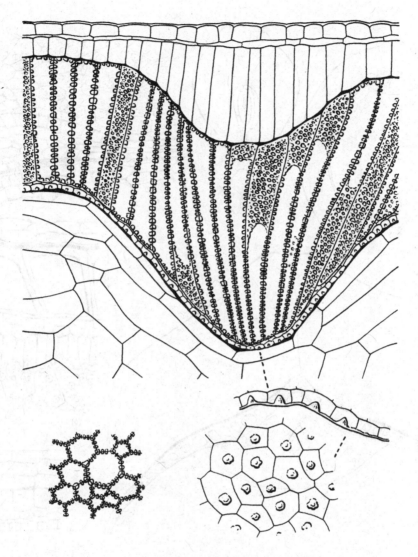

459 *Passiflora edulis.* Testa and tegmen of the fully grown but immature seed in l.s., ×225; exotegmic palisade cells in t.s., ×225; endotegmic cells, ×500.

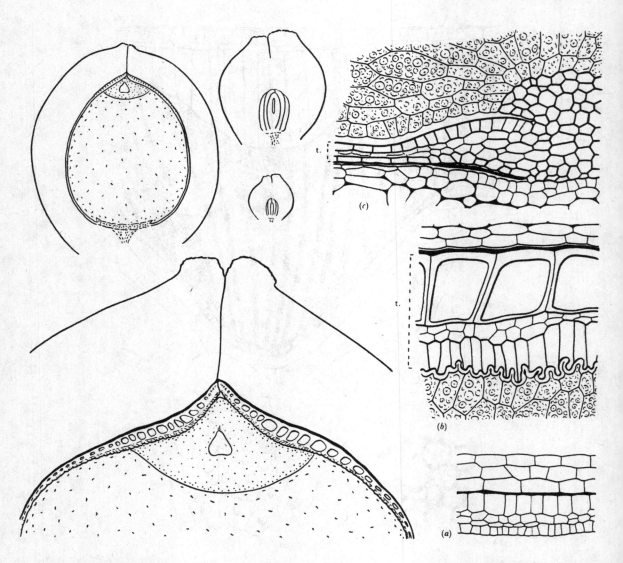

460   *Piper zeylanicum*. Fruit in l.s., × 10. Ovary with ovule in l.s., × 10 and × 25. Apex of fruit and micropylar end of seed with tegmen, endosperm and perisperm, × 50. (*a*) Ovule-wall in l.s. with o.i. and i.i. (lower), × 225. (*b*) Mature seed-coats in l.s. in the micropylar region with adjacent pericarp, tegmen (t.), and perisperm with thick crenulate cuticle and starch-cells, × 225. (*c*) Base of seed in l.s. to show the attachment of the perisperm (starch-cells), tegmen (t.), and pericarp to the dark brown hypostase, × 225.

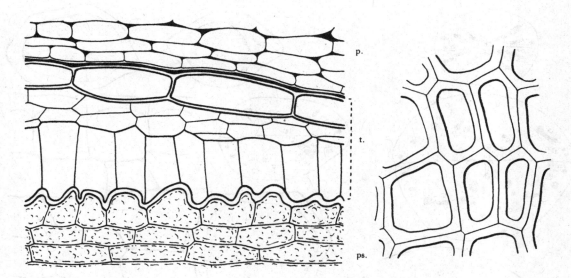

461 *Piper nigrum* (left), micropylar end of the immature seed-coat in l.s. with adjacent pericarp (p.), tegmen (t.) and immature perisperm (ps.) separated by the crenulate cuticle from the endotegmen, × 500. *P. zeylanicum* (right), exotegmic cells of the micropylar region in surface-view, × 225.

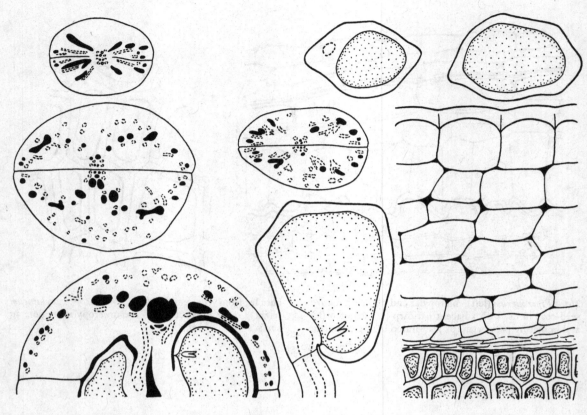

462 *Pittosporum resiniferum.* Seed in l.s. and t.s., × 10. Seed-coat with outer part of the endosperm in t.s., × 225.

Fruit in t.s. from the apex to the middle, × 5. Resin-canals in black.

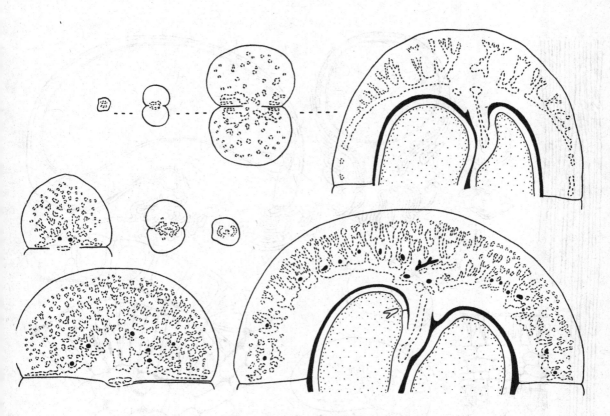

463 *Pittosporum moluccanum* (above) and *P. tobira*
(below). Fruit in t.s. from the style-base to the middle,
× 5. Resin-canals in black.

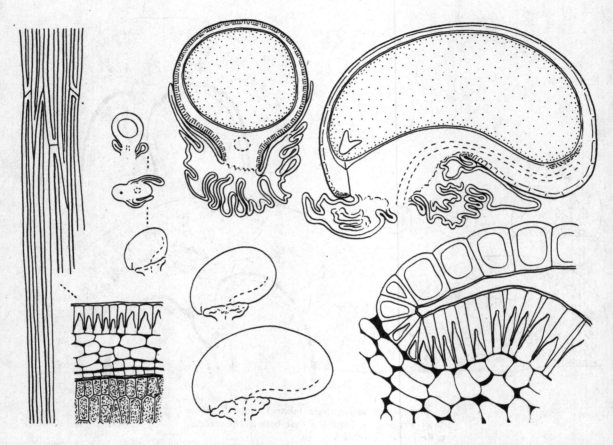

464 *Vancouveria hexandra*. Mature seed in l.s. and t.s., × 20. Developing seed showing micropyle, aril and v.b. of raphe, the youngest in t.s., and basal section across the micropyle, × 25. Mature seed-coat in t.s. with endosperm on the antiraphe-side, epidermal cells in surface-view, and the junction of the aril and raphe in t.s., × 225.

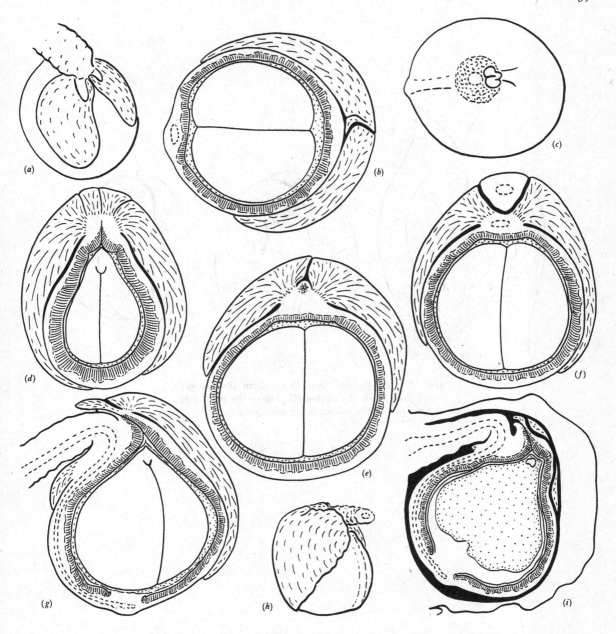

465 *Polygala pulchra.* (*a*) Immature seed with funicle and incipient aril; (*b*) mature seed in t.s.; (*c*) mature seed in micropylar view with the aril removed to show the attachment round the beginning of the raphe (the micropyle intact); (*d*)–(*f*) transmedian l.s. of the seed in the antiraphe half; (*g*) median l.s.; (*h*) mature seed with funicle and aril; (*i*) immature seed inside the fruit-wall. (*a*)–(*g*), (*i*), × 10; (*h*), × 5.

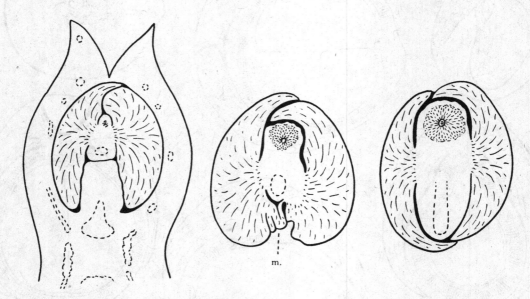

466 *Polygala pulchra*. Seed in t.s. from above downwards (left, inside the fruit-wall) to show the attachment of the aril; m. the micropylar lobes; × 10.

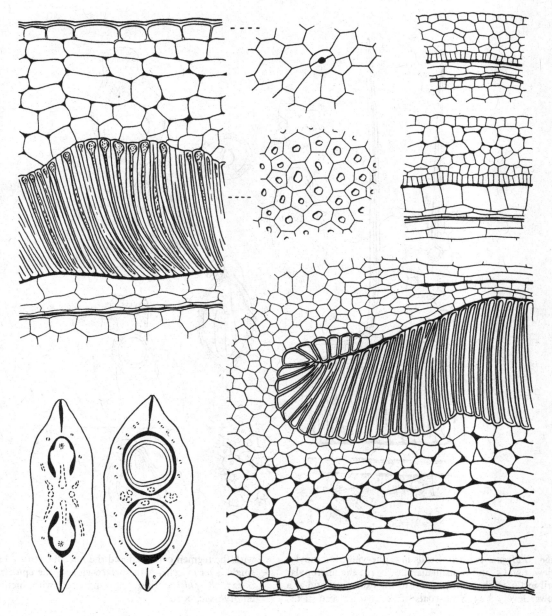

467 *Polygala pulchra*. Wall of ovule and young seed (right) in t.s. and of the mature seed (left) with nucellar remains, ×225. Junction of seed-coats with the chalaza in a fully grown but immature seed, ×225. Young fruit in t.s. to show the inception of the aril, ×10.

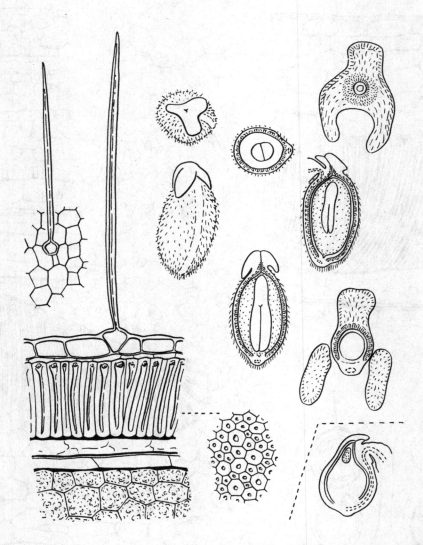

468 *Polygala vulgaris.* Seeds in raphe-view, micropylar view, t.s., l.s. and transmedian l.s., × 10; the micropylar aril in t.s. to show attachment to the exostome, and at a lower level, × 25. Seed-coats of the mature seed in t.s. with the tegmen collapsed and the nucellus reduced to a single layer of cells, with surface-view of the epidermis, × 225. *P. pulchra* (inset); young seed with incipient embryo-sac, × 10.

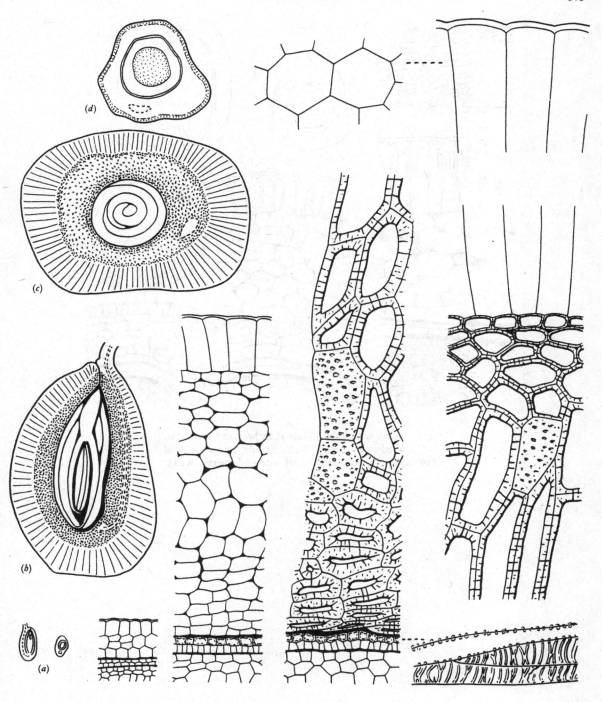

469 *Punica granatum.* (*a*) Ovule in l.s. and t.s., × 10. (*b*) Seed in l.s. with sarcotesta and sclerotic mesotesta, × 5. (*c*) Seed in t.s., × 10. (*d*) Young seed in t.s. with endosperm, nucellus, tegmen and wide testa beginning to form the sarcotesta, × 10. Wall of ovule, young seed and mature seed in t.s., × 225; exotegmic tracheids in surface-view, × 500.

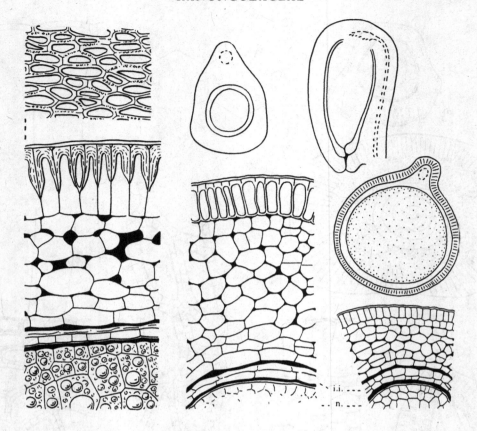

470   *Aquilegia vulgaris*. Ovule at anthesis in t.s. and l.s.,
× 50. Seed in t.s., × 25. Wall of ovule, immature seed
(10 days old), and mature seed with endosperm, × 225.

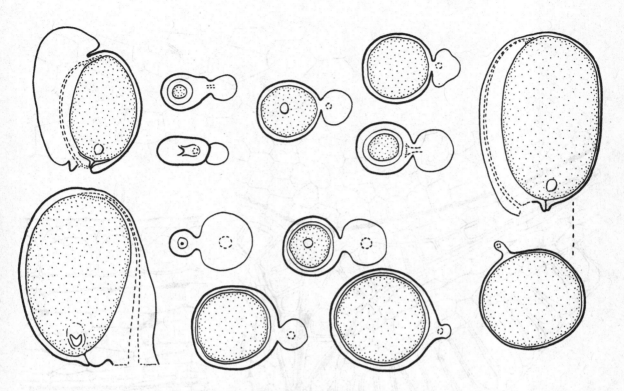

471 *Helleborus corsicus* (upper row), *H. foetidus* (lower row), and *H. niger* (right). Seeds in l.s. and t.s. at various levels from the micropyle, with enlarged raphe, × 10.

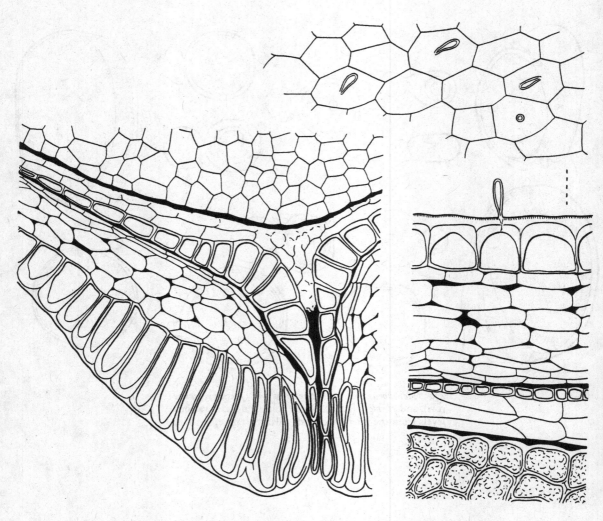

472 *Helleborus corsicus* (left), micropyle of the seed in
l.s., the endosperm immature, × 225. *H. foetidus* (right),
seed-coats with endosperm in t.s., × 225.

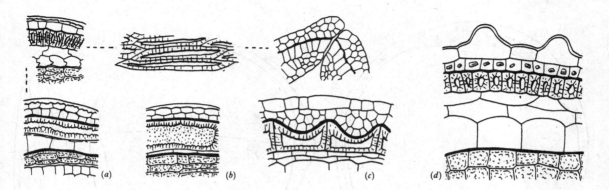

473 *Reseda*, seed-coats with fibrous exotegmen; endosperm shown with cell-contents. (*a*) *R. luteola* in l.s., t.s., and l.s. of the micropyle of a young seed, ×120 (after Crété 1936a). (*b*) *R. lutea*, mag. (after Hennig 1930). (*c*) *R. odorata*, ×100 (after Singh, D. and Gupta 1967). (*d*) *R. alba*, ×225 (after Guignard 1893).

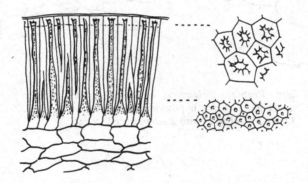

474 *Colletia sp.* (Bolivia). Outer part of the testa in t.s., and sections of the palisade-cells, ×225; exotestal facets, ×500.

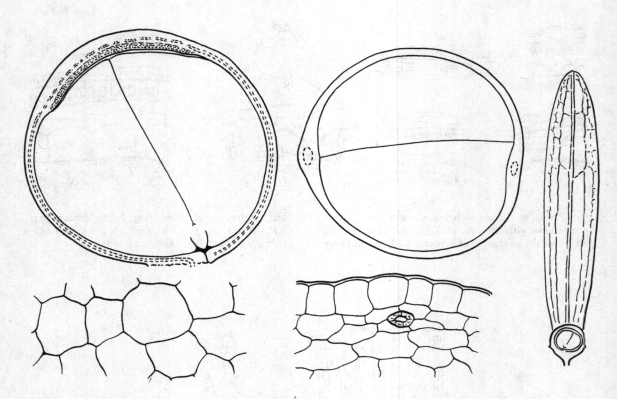

475 *Ventilago malaccensis*. Fruit in l.s., × 1. Mature seeds in l.s. (with hypostase stippled) and t.s., × 12. Outer part of testa with a sclerotic cell, × 225.

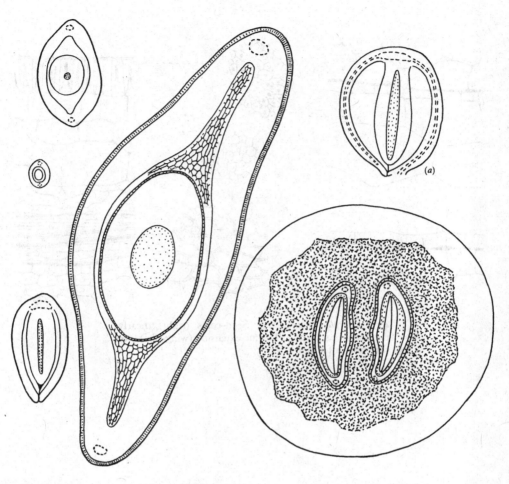

476 *Zizyphus mauritiana*. Ovule in t.s., young seed in t.s. and transmedian l.s., and fully grown but immature seed in t.s., × 18. Ripe fruit in t.s. with woody endocarp, × 5. (*a*) Fully grown but immature seed in l.s., with thick tegmen, × 5.

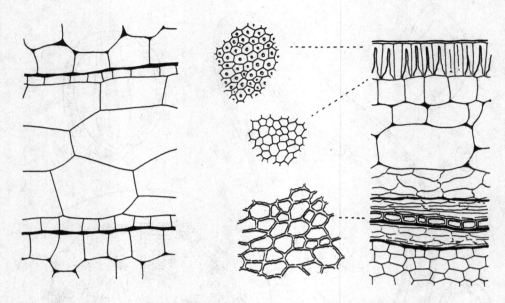

477 *Zizyphus mauritiana*. Seed-coats (right), nearly
mature, in t.s., × 225. Tegmic extension (wing-like) into
the testa in t.s. (left), × 225.

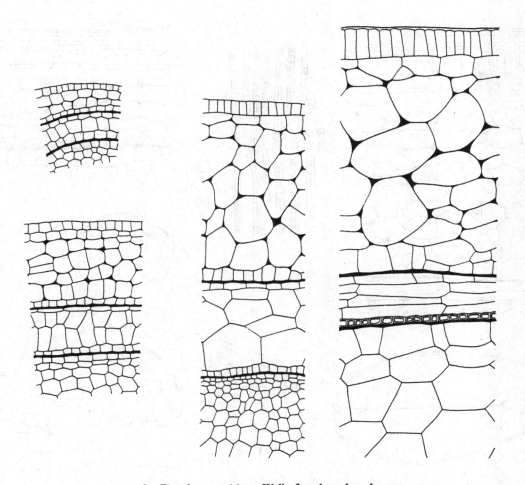

478 *Zizyphus mauritiana.* Wall of ovule and seed-coats
of young seeds in t.s., with nucellar tissue, × 225.

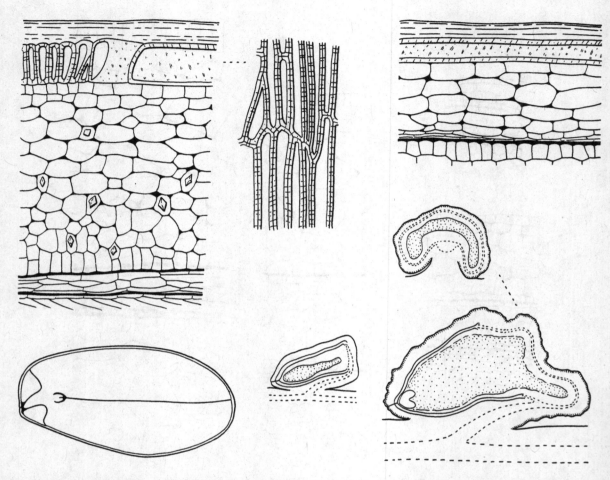

479 *Atalantia monophylla*. Mature seed in l.s., × 5. Young and half-grown (wrinkled) seeds in l.s., and the chalazal end of the older seed in t.s., × 10. Seed-coats of an immature seed (left) and mature seed (right) with gelatinous pellicle, × 225.

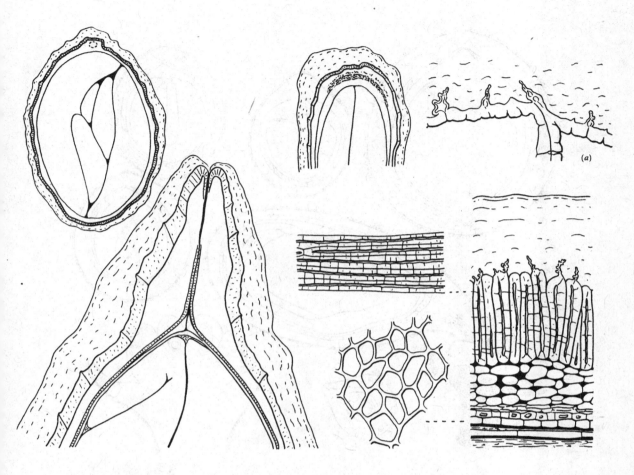

480 *Citrus aurantia*. Seed (with several embryos) in t.s. and the chalazal end of the seed in transmedian l.s., × 10. Micropylar end of the seed in transmedian l.s., the large-celled exotesta with mucilage-pellicle × 25. Seed-coats in t.s. with crushed tegmen and nucellus, × 225. (*a*) Outer wall of an exotestal cell with mucilage-papillae, × 500.

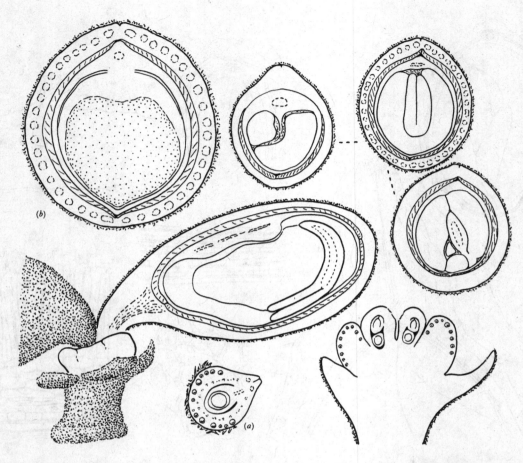

481 *Cusparia sp.* (Manaus, Corner 133). Fruit (nearly ripe) with a follicle in l.s. × 5. Ovary soon after fertilization in l.s., with oil-glands in the disc, × 10. Follicle with seed in t.s. at various levels from the base (left) to the apex, × 5. (*a*) Carpel in t.s., × 25. (*b*) Young follicle with immature seed in t.s., to show the rudimentary tegmen, × 10. Endocarp-cocci hatched.

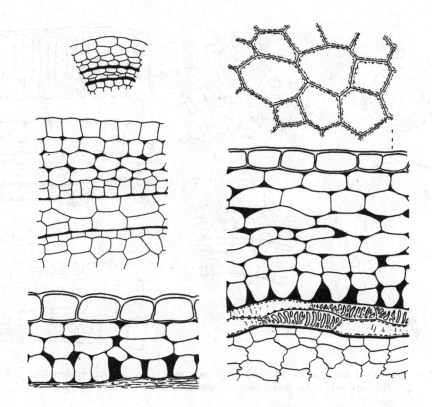

482 *Cusparia sp.* (Manaus, Corner 133). Wall of ovule, young seed, and mature seed in t.s., ×225; mature seed-coat (left) from the antiraphe and (right) from the chalazal region with tracheidal tegmen.

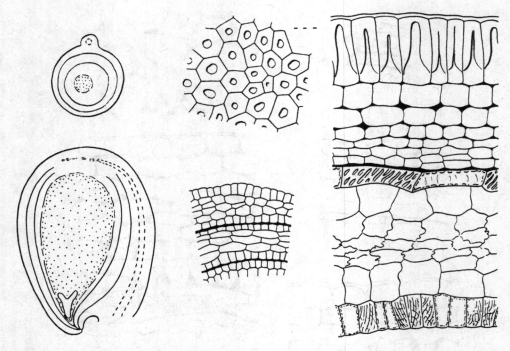

483 *Dictamnus albus*. Immature seeds in l.s. and t.s., with thick tegmen, × 10. Wall of ovule in t.s. and seed- coats in l.s., × 225; exotegmen and endotegmen tracheidal.

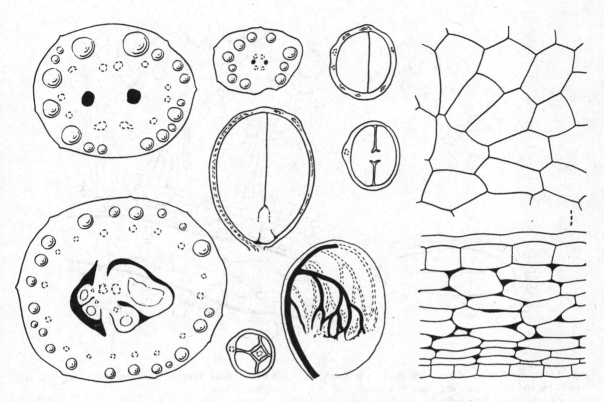

484 *Glycosmis pentaphylla*. Ovary with ovules in t.s., × 10. Style and top of ovary in t.s., × 25. Seed in l.s. and in t.s. at various levels, × 5. Testa in l.s., × 225. Vascular supply of the seed, × 5.

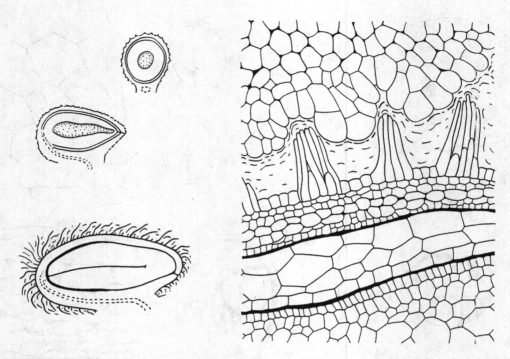

485   *Limonia acidissima*. Mature seed in l.s., × 8. Young seed in l.s. and t.s., × 10. Seed-coats of a young seed in t.s., with incipient hair-tufts in mucilage, with the placental tissue bulging between them, tegmen and nucellus, × 225.

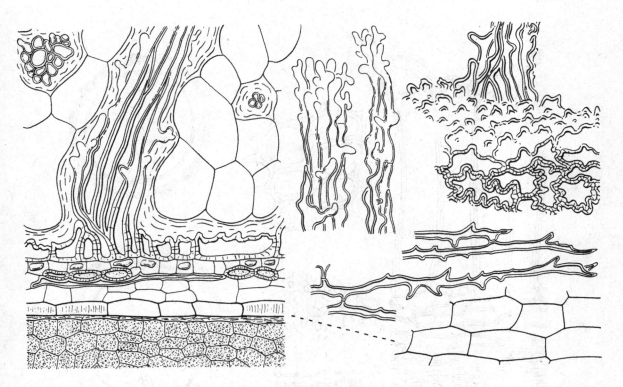

486 *Limonia acidissima*. Nearly mature seed-coat in t.s. with a hair-tuft entering the fruit-pulp, and two hair-tufts in t.s., tegmen and nucellus crushed, the testa resting on the endosperm, ×225. Tips of two hair-tufts, exotesta with the base of a hair-tuft, fibres of the meso-testa, and the endotesta, ×225.

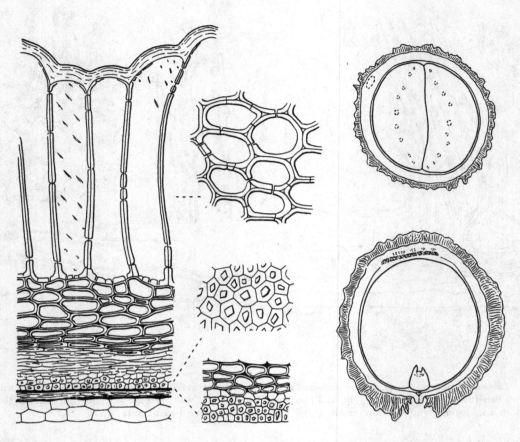

487 *Murraya glenei*. Seeds in t.s. and l.s., × 10. Seed-coats in t.s., with exotestal palisade covered by a mucilage-pellicle, inner mesotestal tissue collapsed, tegmen and nucellus crushed, endosperm 2 cells thick, × 225; part of the inner testa before collapse, × 225; endotestal facets, × 500.

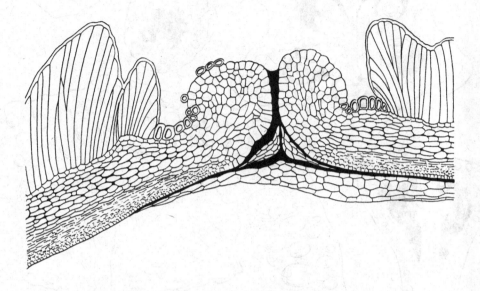

488 *Murraya glenei*. Micropyle of mature seed in l.s.
with a trace of nucellus, × 110.

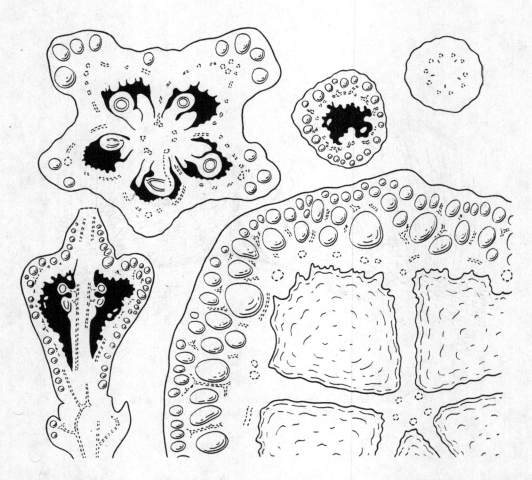

489 *Murraya glenei*. Ovary in l.s. soon after fertilization, with endocarp papillae, × 10. Similar ovary in t.s. across the shoulders, at the apex, and across the style (with 5 slit-like stylar canals), × 25. Fruit in t.s., above the seeds, with cortical v.b. between the oil-glands, × 8.

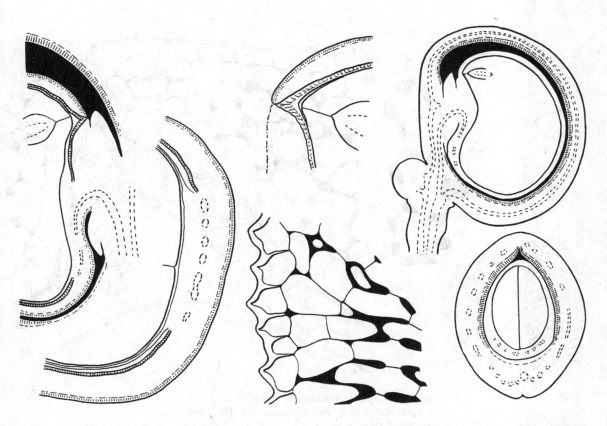

490 *Pilocarpus racemosus*. Fruit in l.s. with the one developed drupe, and the drupe in t.s. (endocarp hatched), × 5. Micropyle in l.s. and chalazal part of the seed in t.s., with exotesta and exotegmen hatched, × 25. Placenta, funicle, aril and seed in l.s., with endocarp, exotesta and exotegmen hatched, × 10. Epidermis of aril, × 225.

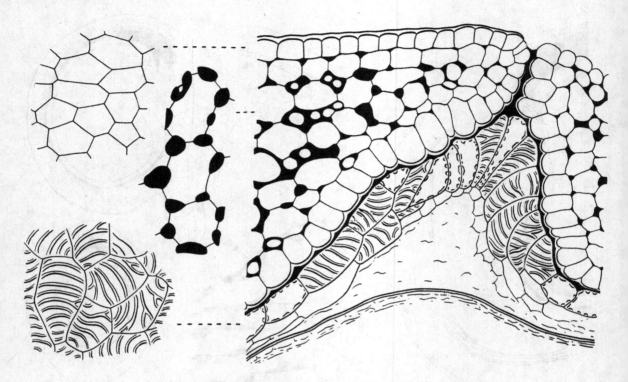

491 *Pilocarpus racemosus.* Micropyle of seed in l.s. to the nucellar cuticle and remains, with the tissues in tangential view, × 225.

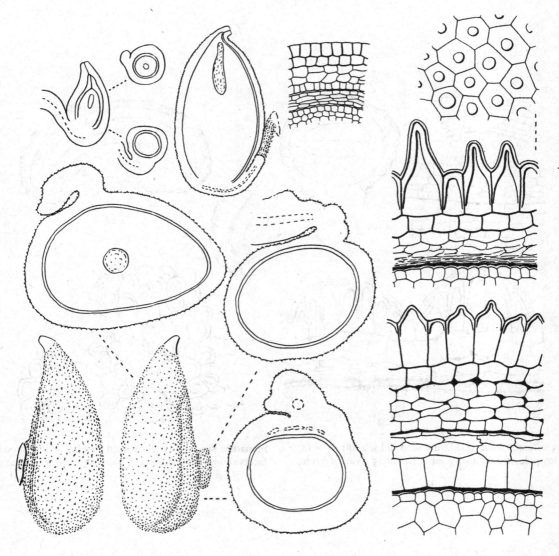

492 *Ptelea trifoliata.* Immature but full-sized seeds in side-view and l.s., ×10; seeds in t.s. at three levels to show the curved funicle and the extension along the preraphe, ×25. Ovule in l.s. and t.s., ×25. Ovule-wall, immature and nearly mature seed-coats with the periphery of the nucellus in t.s., ×225.

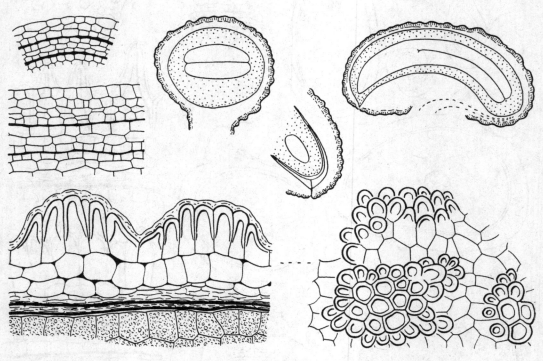

493 *Ruta graveolens*. Mature seed in l.s. and t.s., × 18. Micropyle of immature seed in l.s., × 25. Wall of ovule, immature seed and mature seed in t.s., with exotestal facets showing the clusters of more elongate cells, × 225.

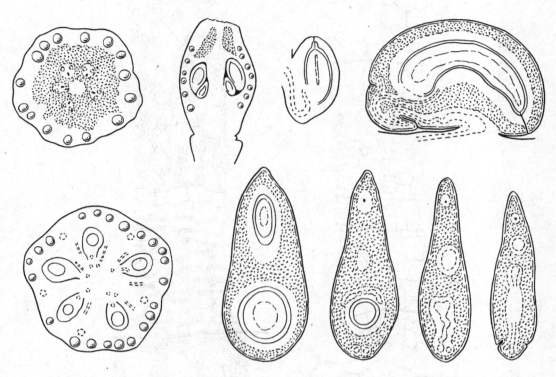

494 *Toddalia asiatica*. Immature seed in l.s. and transmedian l.s. at four levels from the micropyle (lower right) to the middle, showing the curved raphe and sclerotic mesotesta (speckled), × 10. Ovule in l.s., × 25.

Young fruit in l.s., with tannin-cells stippled, × 10. Young fruit in t.s. in the region of the tannin-cells and across the ovules, × 25.

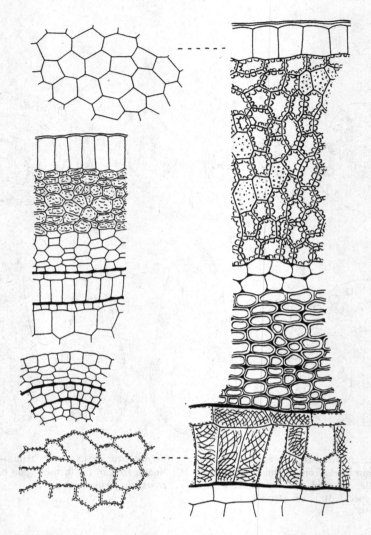

495 *Toddalia asiatica*. Wall of ovule, young seed and nearly mature seed in t.s., with nucellar tissue, ×225; mesotesta sclerotic in the outer part derived from the tanniniferous layer of the young seed, and the tegmen becoming tracheidal.

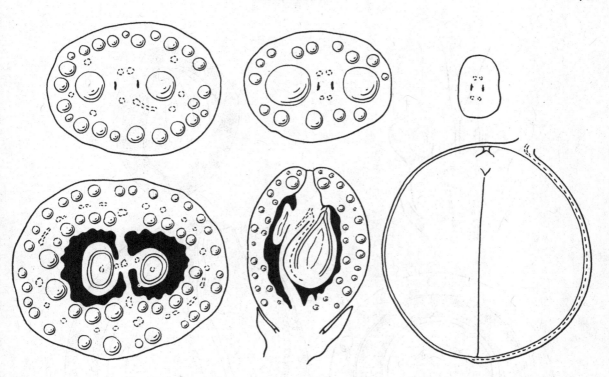

496 *Triphasia aurantia*. Seed in l.s., ×8. Ovary soon after fertilization in l.s., with an abortive ovule, ×8.

Style, upper part of the fruit and placental region in t.s., ×25.

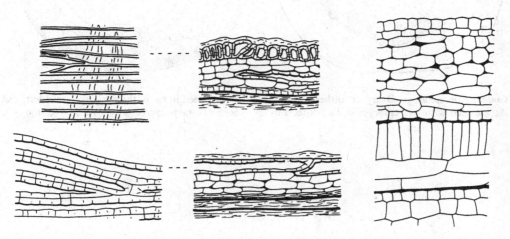

497 *Triphasia aurantia*. Seed-coats of young seed with nucellar tissue (right) in t.s., and of the mature seed with mucilage-pellicle in t.s. and l.s., ×225; exotestal fibres (lower left), ×500.

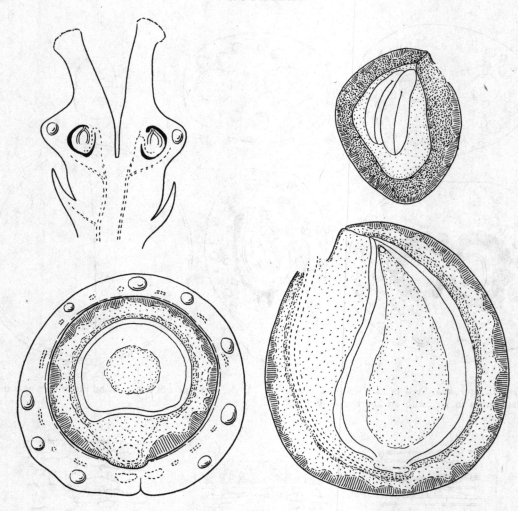

498 *Xanthoxylum simulans*. Ovary at anthesis in l.s., × 18. Mature seed with two embryos in l.s., × 10. Im- mature seed in t.s. in the pericarp (lower left) and in l.s., × 18. Testa 3-layered (as shown in Fig. 499).

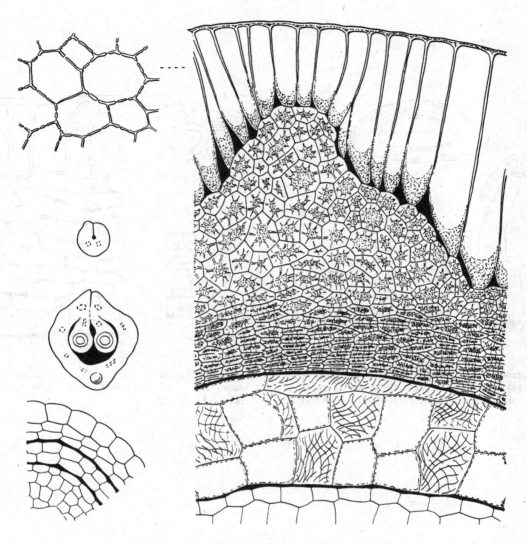

499 *Xanthoxylum simulans.* Seed-coats of nearly mature
seed with nucellar tissue, × 225. Ovule-wall in t.s., × 225.
Style and carpel in t.s., × 25.

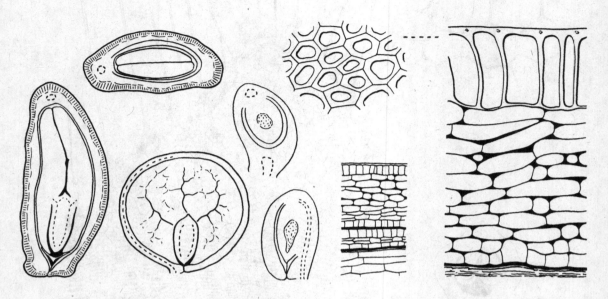

500 *Azima tetracantha*. Mature seeds in transmedian l.s. (×12) and median l.s. (×6), the orbicular cotyledon with impressed veins. Immature seed in t.s. with tegmen and nucellar remains, ×12. Ovules in median and transmedian l.s., ×25. Ovule-wall in t.s. and seed-coat (testa) in l.s., ×225.

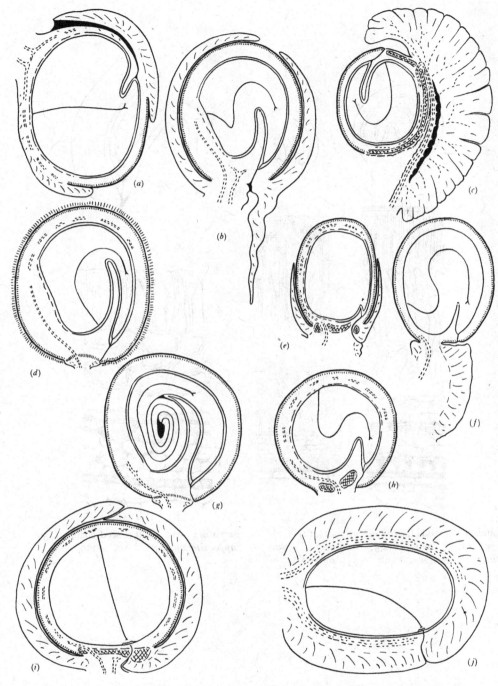

501 Sapindaceous seeds, semidiagrammatic, not to scale, with internal structure so far as known; sclerotic hilum hatched, pachychalaza as a broken line, exotesta closely striated, aril widely striated. (a) *Harpullia sp.* (New Guinea), with partial sarcotesta along the raphe and long micropyle. (b) *Guioa pubescens.* (c) *Alectryon sp.* (New Guinea), with raphe-aril. (d) *Harpullia zanguebarica,* with hairy testa and vestigial aril. (e) *Cupania sp.* (Brazil). (f) *Trigonachras,* with funicular aril. (g) *Dodonaea viscosa,* with vestigial aril. (h) *Allophylus,* exarillate. (i) *Euphoria malaiensis.* (j) *Nephelium lappaceum,* with sarcotesta.

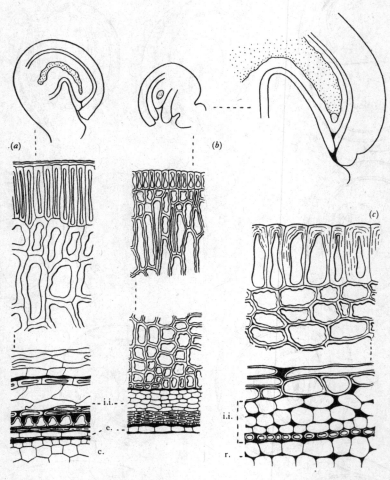

502 Sapindaceous seeds (after Guérin 1901). (*a*) *Koel-reuteria paniculata*, ovule ×25; seed-coats ×250. (*b*) *Xanthoceras sorbifolia*, ovule ×35; micropyle of immature seed ×25; seed-coats ×150. (*c*) *Aesculus hippocastanum*, seed-coats, ×150.

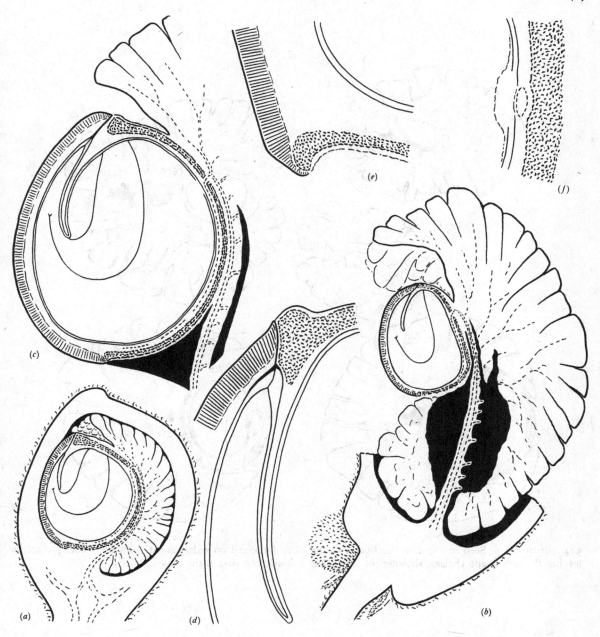

503 *Alectryon sp.* (*a*) Seed in the fruiting carpel and (*b*) the seed elevated from the carpel (transversely dehisced) by the enlargement of the aril, × 5. (*c*) Seed in median l.s., × 10. (*d*) Micropyle and radicle-pocket, × 25. (*e*) Junction of the testal palisade with the sclerotic layer at the lower end of the seed (cf. (*c*), × 25. (*f*) Chalaza in t.s. with the origin of the tegmen, the remains of the nucellus, and the raphe v.b. in the inner tissue of the testa, × 25. Exotestal palisade hatched, sclerotic layer of the testa stippled.

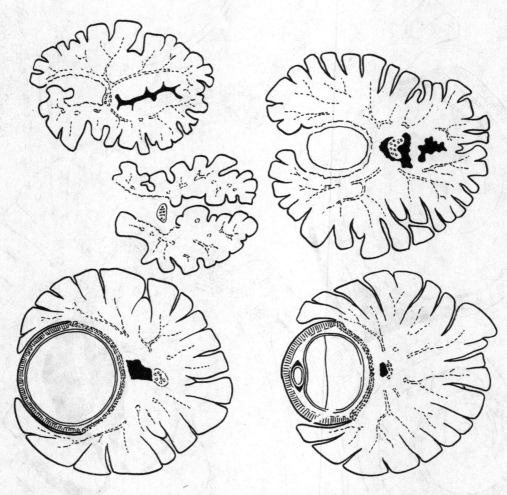

504 *Alectryon sp.* Seed in t.s. from the funicle to the level of the radicle and chalaza, showing the separation of the aril on enlargement from the funicle, ×5; the sclerotic part of the testa stippled.

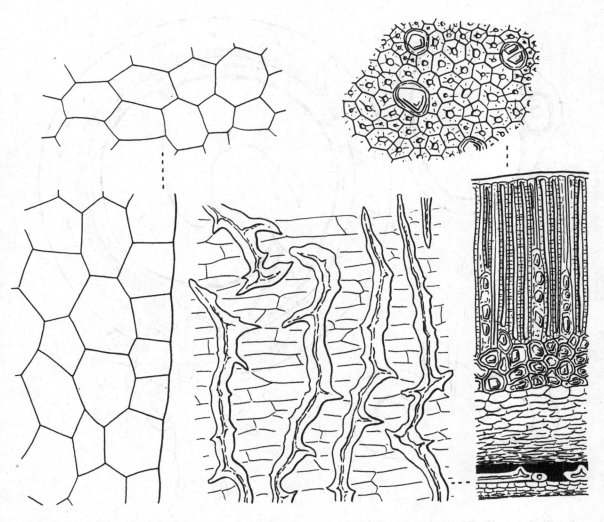

505 *Alectryon sp.* Aril (left) in section at the surface and with epidermal facets, × 120. Testa (right) and tegmen in l.s., × 225. Fibres of the tegmen (centre) resting on the underlying o.h. (tegmen), × 225.

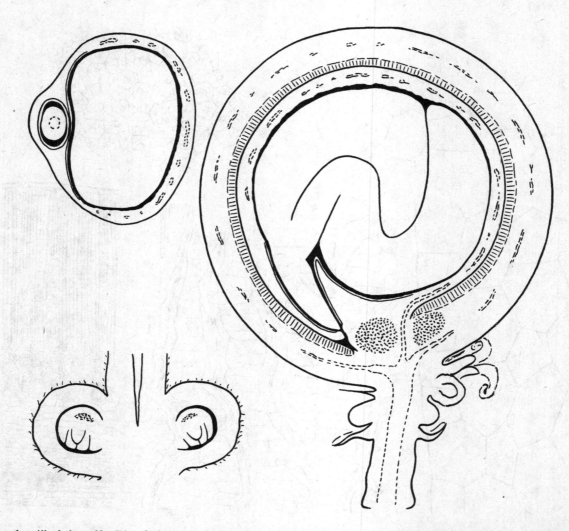

506 *Allophylus cobbe*. Ripe fruit in l.s., with endocarp (hatched) and sclerotic ring in the hilum (speckled), ×10. Seed in t.s. across the radicle, showing the pachy- chalazal structure with the tegmen limited to the radicle-pocket and the crescentic ridge, ×8. Ovary in l.s., with hypostase in the ovules, ×25.

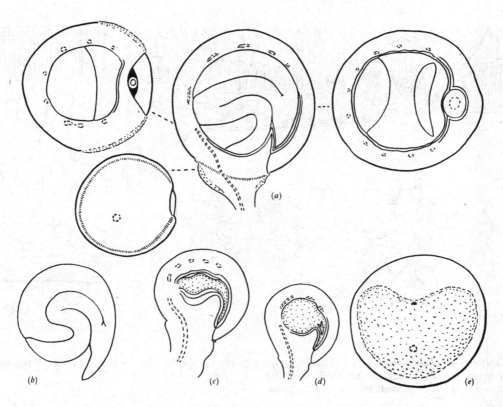

507 *Cardiospermum halicacabum.* (*a*) Mature seed in l.s. and t.s. (at three levels) to show the pachychalazal construction, the hilar disc or palisade, and the funicular aril, × 10. (*b*) Embryo in l.s., × 10. (*c*) Ovule soon after fertilization with incipient aril, × 25. (*d*) Young seed with pachychalazal growth, × 10. (*e*) Mature seed detached from the funicular aril, showing the white heart-shaped hilum with v.b. and micropyle, × 10.

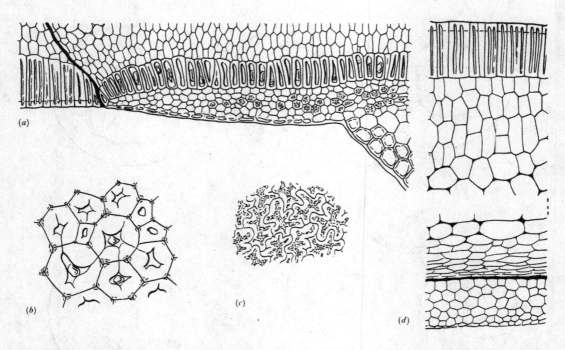

508 *Cardiospermum halicacabum*: (*a*) l.s. of the micropylar slit (excentric) with the adjacent testa, aril, and hilar palisade, × 225; (*b*) hilar palisade in t.s., × 500; (*c*) exo-testa in surface-view, × 500; (*d*) testa and tegmen in t.s., × 225.

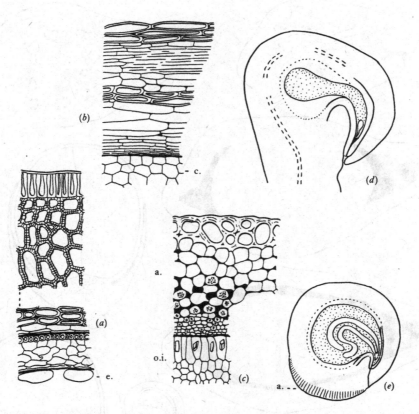

509 *Cardiospermum halicacabum* (after Guérin 1901). (*a*) Seed-coats in t.s. of the micropylar region, of the chalazal region (*b*), and the arillar region (*c*), ×150. (*d*) Young seed in l.s., ×36. (*e*) Older seed in l.s., with arillar region (*a*), ×9.

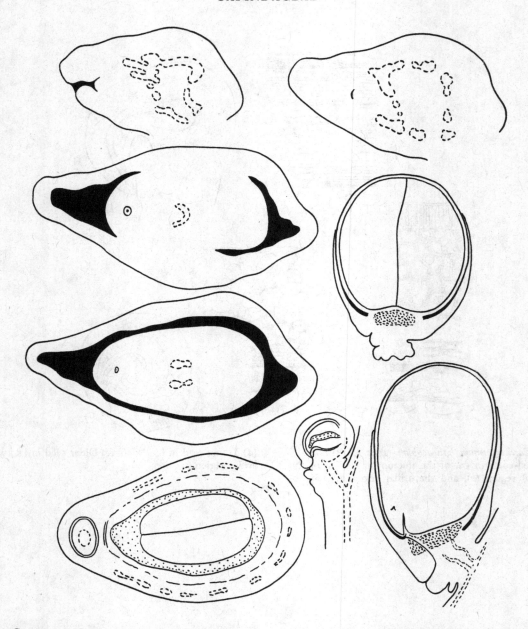

510 *Cupania sp.* (Corner 254). Mature seed with aril in median and transmedian l.s., ×5. Ovule soon after fertilization, with incipient aril and short thick funicle, ×25. Immature seed in t.s. from the funicle (with arillostome) to the level of the radicle-pocket, to show the vasculature, the insertion of the aril, and the thick nucellus (lower figure with endosperm and embryo), ×18.

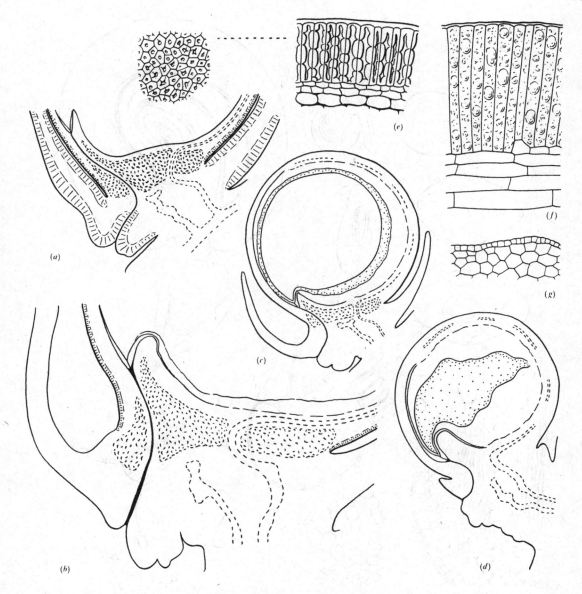

511 *Cupania sp.* (Corner 254). (*a*) Base of the mature seed in l.s. with fully developed aril, the sclerotic tissue in the hilum hatched, × 10. (*b*) Similar section of an immature seed (as (*c*)) to show the vestigial tegmen and immature aril without cortex, × 25. (*c*) Immature seed with endosperm and remains of the nucellus, × 10. (*d*) Very young seed with distinct tegmen, incipient aril, incipient endosperm, and thick nucellus, × 25. (*e*) Exotestal palisade, × 225. (*f*) Cortex of the mature aril in l.s., × 225. (*g*) Outside of the immature aril in t.s., × 225.

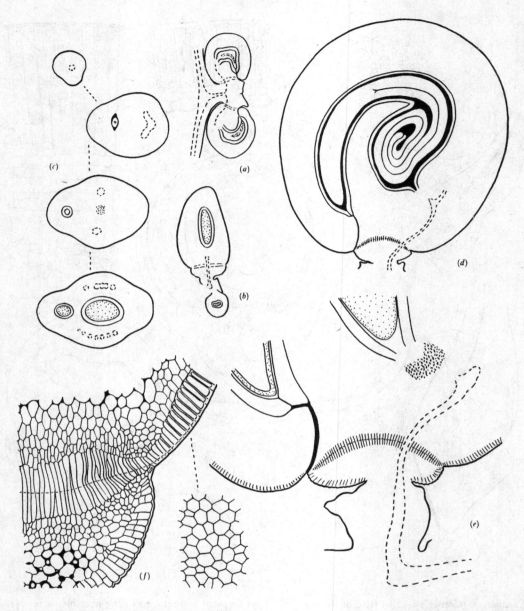

512 *Dodonaea viscosa.* (*a*) Two ovules attached to the placental projection, soon after fertilization, × 25. (*b*) Immature seed in transmedian l.s., with undeveloped ovule in t.s., × 10. (*c*) Four transverse sections of the young seeds from the funicle to the middle part of the seed, to show the vascular supply, × 10. (*d*) Mature seed in l.s., to show the funicular aril in the hilum, × 18. (*e*) The hilar region of a fully grown but immature seed in l.s., with micropyle, endostome, trace of nucellus, chalaza, hypostase, and hilar palisade, × 50. (*f*) Part of (*e*) enlarged (× 225); exotestal facets, × 500.

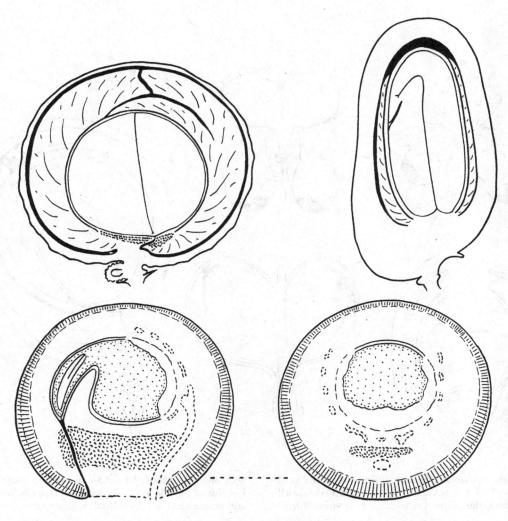

513 *Euphoria malaiensis* (upper left), fruit in l.s. with thick exostomal aril, thin testa, and sclerotic layer in the hilum, × 3. *Pometia pinnata* (upper right), fruit in l.s. with aril, × 2. *Sapindus mukurossi* (lower figures), im- mature seed in l.s. and oblique t.s. showing the pachy- chalaza with v.b., the sclerotic hilum, and the unevenly developed exotestal palisade with *linea lucida*, × 4.

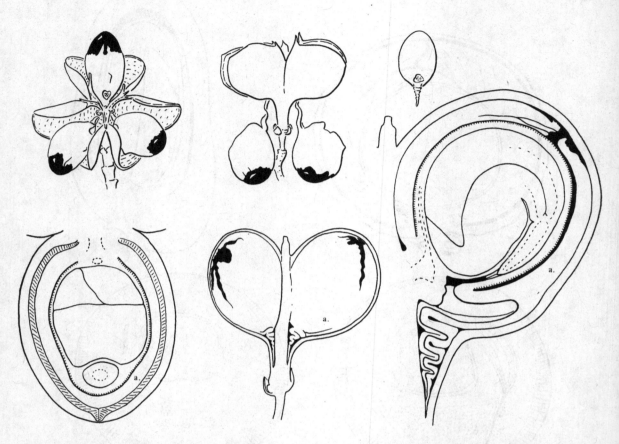

514 *Guioa pubescens.* Dehisced fruits and arillate seed (hilar view), ×2. Loculus of the capsule in t.s. with woody cocci, the exotesta striated, ×7. Unopened fruit with one side removed to show the arillate seeds, ×3. Loculus of capsule in l.s., showing the true (micropylar) arillostome and the false (distal) arillostome, ×7. a. aril.

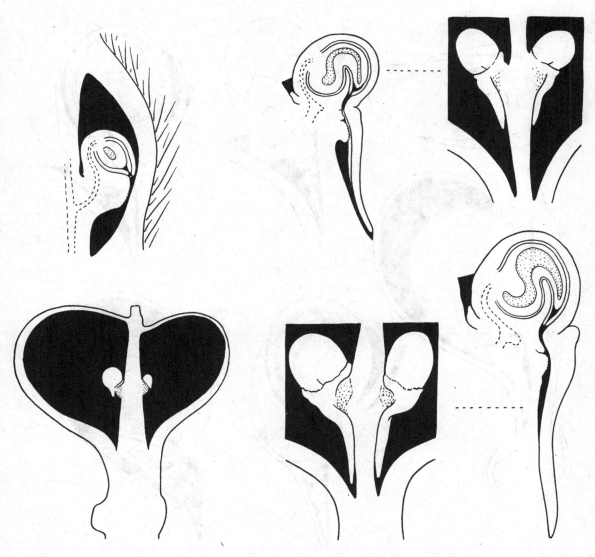

515 *Guioa pubescens*. Part of the ovary at anthesis in l.s. with one ovule, × 50. Developing fruits and seeds to show the lengthening micropylar aril and its cupular upgrowth from the funicle, × 7; young seeds in l.s., becoming campylotropous, × 7.

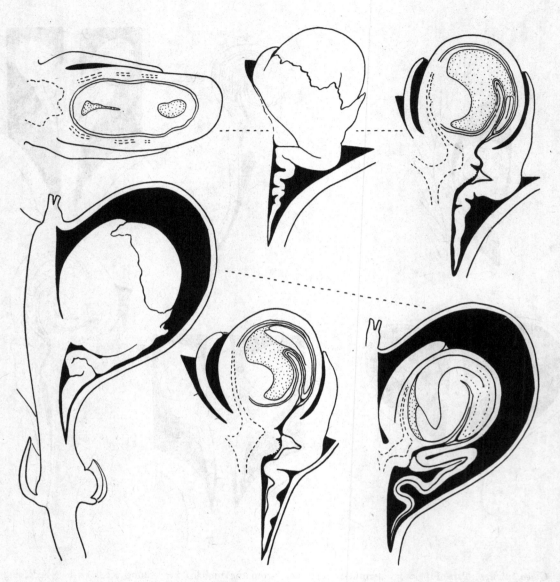

516 *Guioa pubescens.* Development of older seeds, ×7;
half-grown seed in t.s. (upper left), ×15.

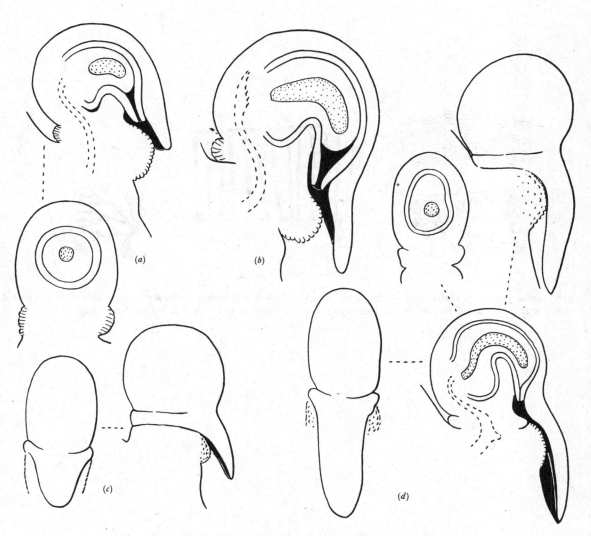

517 *Guioa pubescens*. Stages in the early development of the seed; (*a*), (*b*) soon after fertilisation, becoming campylotropous, × 50; (*c*), (*d*), × 30.

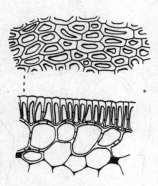 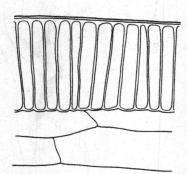 

518 *Guioa pubescens*. Outer part of the testa in t.s. (left), × 225. Surface of the aril in t.s. (centre), × 225. Vascular supply of the seed on one side, with the partial septum of the seed in broken lines, mag.

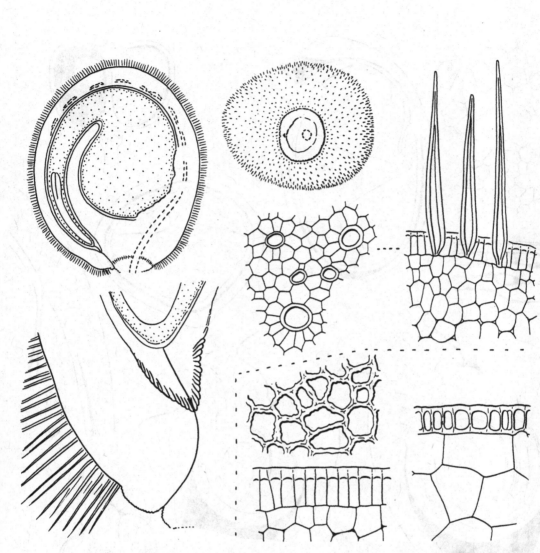

519 *Harpullia zanguebarica*. Immature but fully grown seed in l.s. and micropylar view, × 8; the pachychalaza slightly developed, the aril as a vestige at the hilum. Micropyle in l.s. showing the sclerotic endostome, × 50. Outer part of the testa, × 225; facets, × 500. *Harpullia sp.* (New Guinea), inset; surface of testa and aril (right), × 225; exotestal facets, × 500.

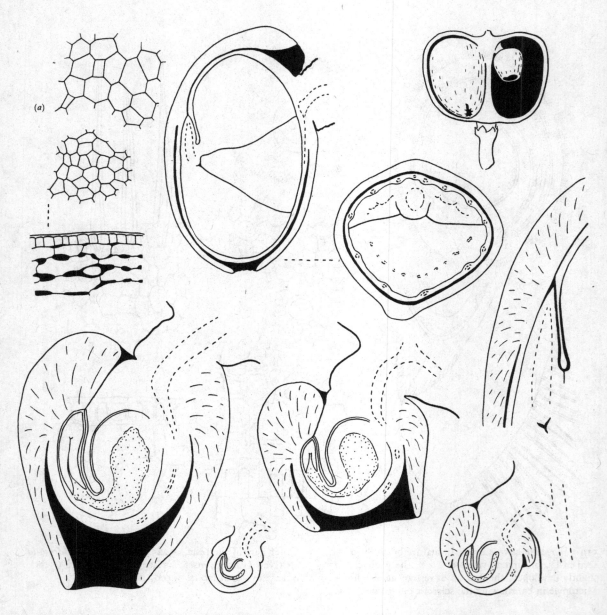

520 *Harpullia sp.* (Solomon Islands). Capsule with the valves removed to show a mature seed and an immature, × 1. Mature seed in l.s. and t.s., covered with the aril, × 3. Developing seeds in l.s., with the tegmen limited to the antiraphe side, the aril striated, × 8. Radicle-pocket of the mature seed, with part of the aril, × 8. Testa in t.s. of the outer part, × 225. (*a*) Facets of the aril-epidermis, × 225.

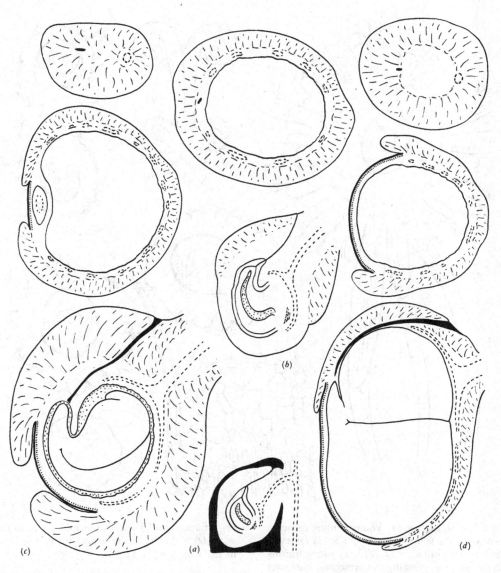

521 *Harpullia sp.* (New Guinea). Immature seed in 5 transverse sections to show the attachment of the arillar tissue (hatched), the arillostome leading to the micropyle (as in (d)), and the v.b., ×5. (a) Ovule in the ovary-loculus, ×25; (b) very young seed showing the enlarging pachychalaza and the incipient aril, ×25; (c) immature seed with fully developed aril, disintegrated tegmen, large pachychalaza, and the remains of the endosperm, ×10; (d) mature seed in l.s., showing the long arcuate arillostome, the wide pachychalaza, the short radicle-pocket, and the free testa, ×5.

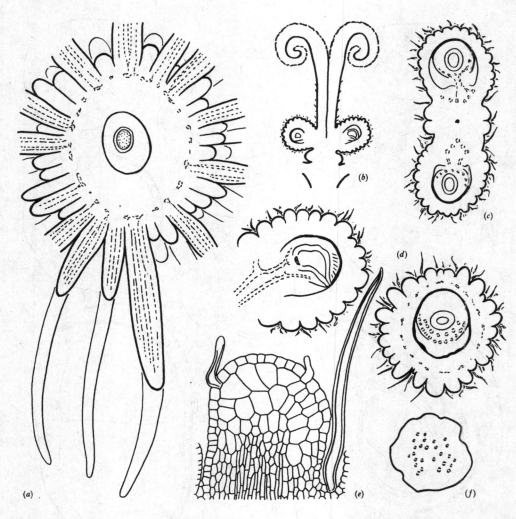

522 *Nephelium lappaceum*. (*a*) Young fruiting carpel in t.s. with most of the spines cut short, ×8. (*b*) Ovary in l.s., ×10. (*c*) Ovary in t.s., ×25. (*d*) Very young fruiting carpel in l.s. and t.s., showing the crescentic obturator (as a funicular aril) and the incipient v.b. to the testa, ×25. (*e*) Tubercle of the ovary, ×225. (*f*) Fruit-spine in t.s. with v.b., ×25.

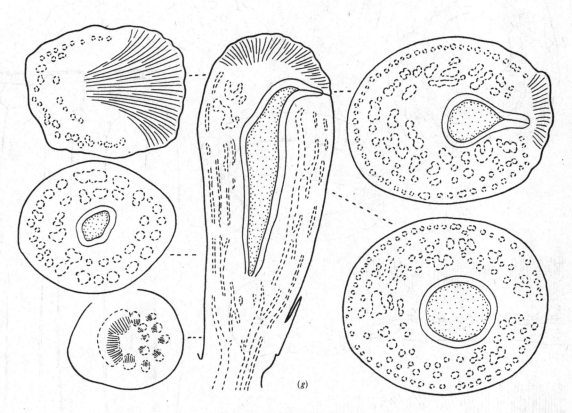

522 *Nephelium lappaceum* (g) Young seed (embyo as yet microscopic) in l.s. to show the extended sides of the ovule with v.b. and the antiraphe with radiating rows of cells without v.b., × 10. Transverse sections of parts of a similar seed, × 25.

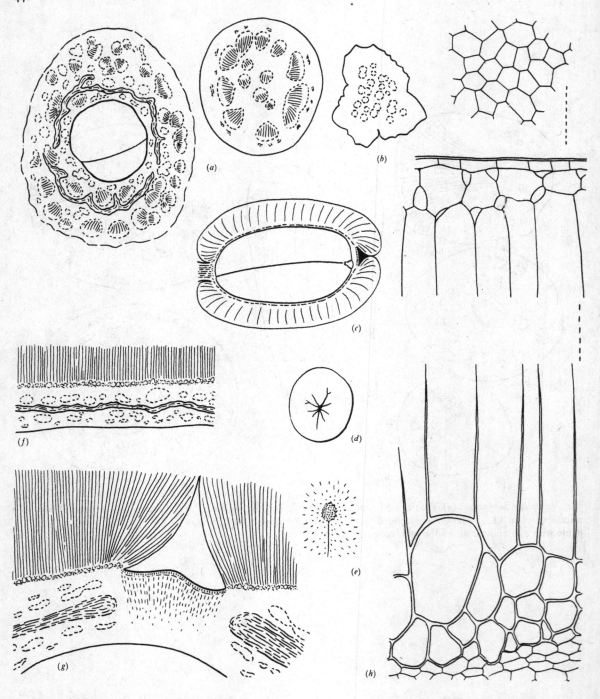

523 *Nephelium lappaceum*, structure of the ripe seed: (*a*) funicle and base of seed in t.s. at the junction of funicle and sarcotesta, to show v.b. and sclerotic layer, ×10; (*b*) fruit-spine in t.s. near the base, with small fibrous v.b., ×25; (*c*) ripe seed in l.s. with funicle, sarcotesta, embryo, and micropylar area, ×2; (*d*) seed in micropylar view, the folds of the sarcotesta leading to the micropylar area, ×1; (*e*) the micropylar area with the sarcotesta removed, ×2; (*f*) junction of the sarcotesta and inner testa in t.s., showing the v.b. and sclerotic layer, ×25; (*g*) micropylar area in t.s., ×25; (*h*) sarcotestal tissue in section, ×225.

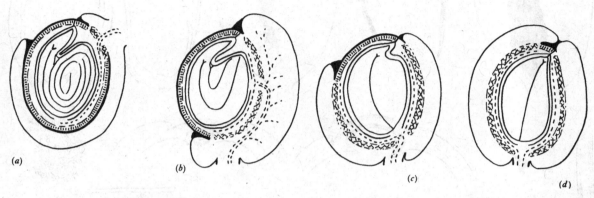

524 Diagrams to show the possible derivation of the sarcotestal seed of *Nephelium* (*d*) from an exotestal seed with funicular aril (*a*) through expansion of the pachy- chalaza as in *Alectryon* (*b*), with straightening of the embryo (*c*); exotesta striated, sclerotic tissue of the pachychalaza hatched.

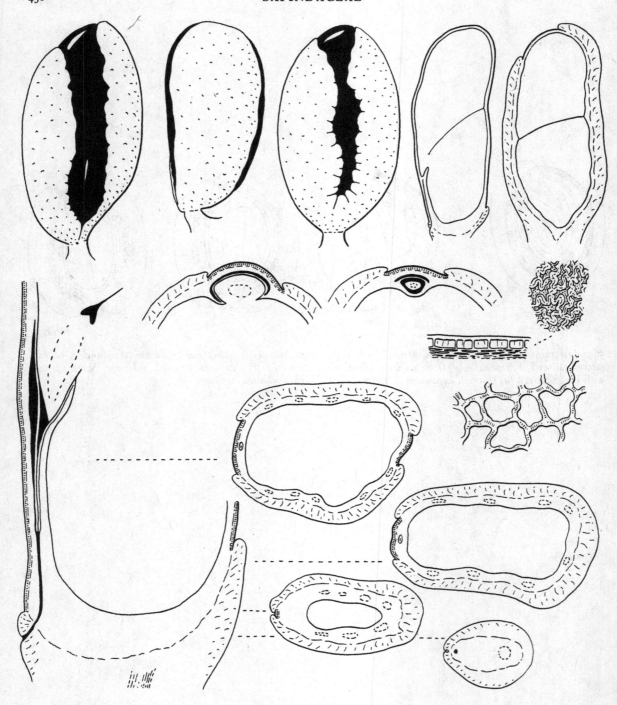

525 *Paullinia pinnata*. Seed (from left to right) in micropylar, lateral and raphe-view, in ·median and transmedian l.s., × 5. Endostome and exostome, with radicle and minute plumule, in l.s., × 18. Basal part of the seed in t.s. at various levels, × 10. Testa in section and exotestal facets, with the epidermal facets of the aril, × 225. Aril and its adnate sarcotestal part hatched.

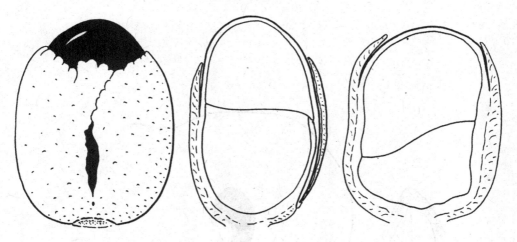

526 *Paullinia riparia*. Seed in micropylar view and in
median and transmedian l.s., × 5.

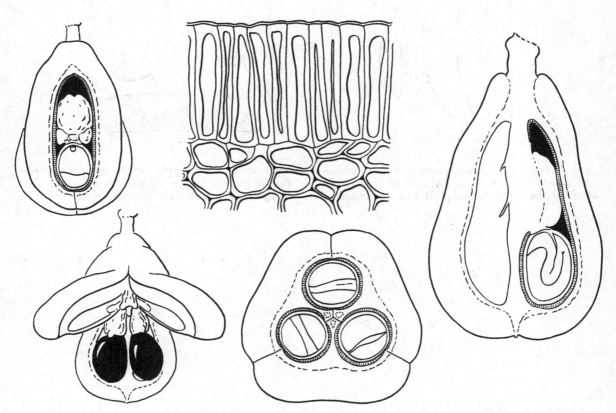

527 *Trigonachras acuta*. Left, dehisced fruit and tangential section to show the seed with funicular aril in the
loculus, × 1. Fruit in l.s. and t.s. with the cocci striated
and the woody inner mesocarp marked by a broken line,
× 1½. Testa in t.s. of the outer part, × 400.

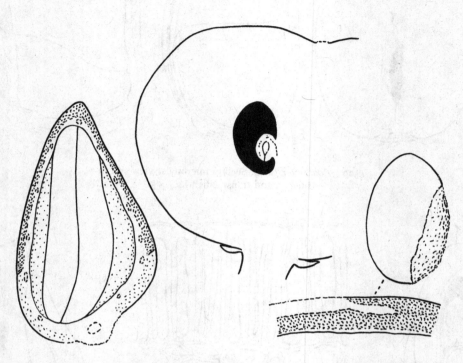

528 *Chrysophyllum cainito*; young fruit in ls. to show the growth of the loculus, ×5; seed in t.s. with the lignified tissue (heavily stippled) and the unlignified outer tissue (lightly stippled) of the testa, ×5. *C. roxburghii* (right); seed, ×1; junction of woody testa (stippled) in t.s. with the hilar region, ×10.

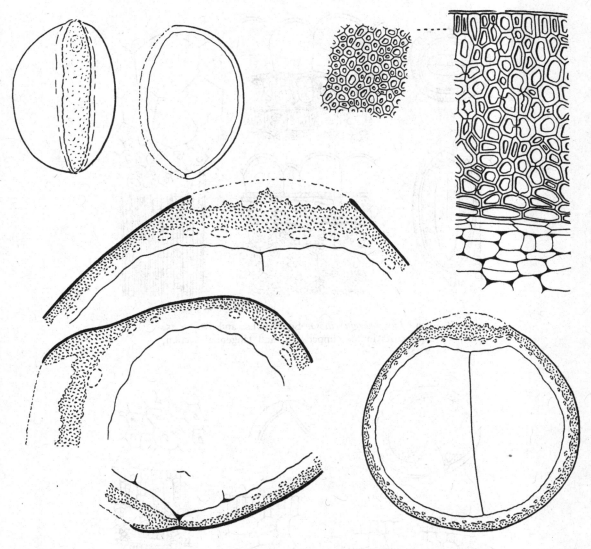

529 *Planchonella dulcifica*. Seed in hilar view and l.s., × 2. Seed in t.s., × 5; placental region in t.s., and chalazal and micropylar regions in l.s., to show the woody tissue (stippled) of the testa and the v.b., × 10. Testa in t.s. to show the woody outer tissue and the thin-walled inner tissue, × 225.

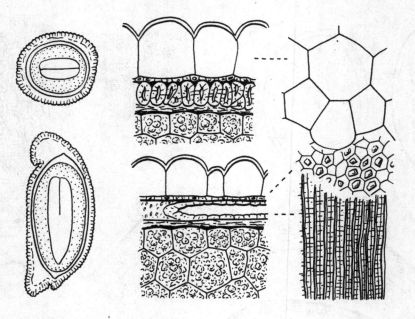

530 *Luxemburgia ciliosa.* Seed in l.s. and t.s., × 25. Seed-coats in t.s. (upper), l.s. and tangential section, × 225.

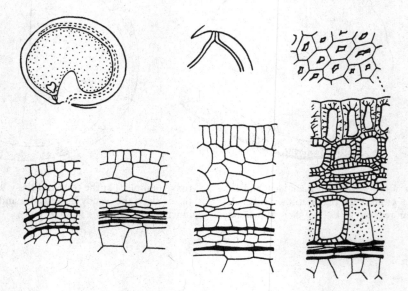

531 *Schisandra grandiflora* (after Kapil and Jalan 1964). Seed in l.s., × 2. Micropyle of ovule, × 30. Wall of ovule and of developing seeds in t.s., × 160.

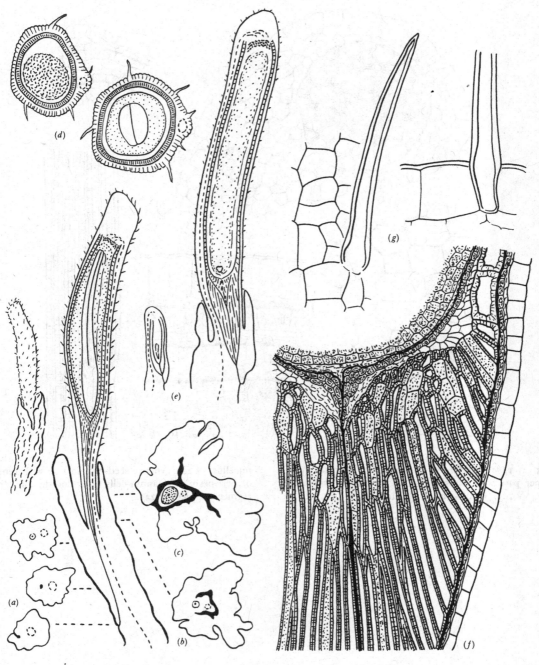

532 *Scyphostegia borneensis*. Mature seed with aril, × 4. Mature seed in median l.s., with t.s. at different levels, × 10: (*a*) three sections of the pseudofunicle in the part of the elongated exostome, × 10; (*b*) section at the separation of the aril, × 15; (*c*) section of the upper part of the pseudofunicle in the region of the elongated endostome, and the surrounding aril, × 25; (*d*) sections of the coty-ledonary part of the seed with trace of perisperm, and at the level of the hypostase (brown tissue, stippled), × 25; (*e*) ovule and young seed in median l.s., × 25; (*f*) base of woody endostome in l.s., with adjacent endosperm and testa, the tannin-cells of the endotesta and endotegmen stippled, × 120; (*g*) hairs of the testa, × 225.

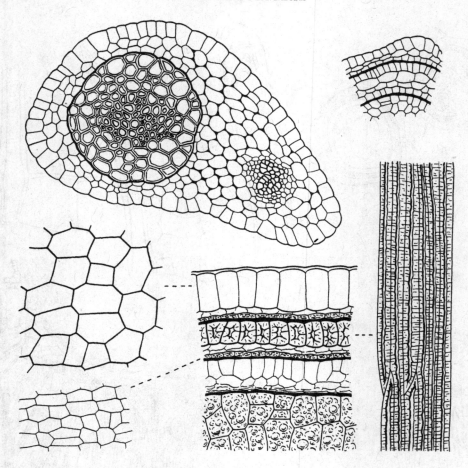

533   *Scyphostegia borneensis.* Pseudofunicle in t.s. in the upper part with massive woody endostome and small v.b., × 225. Ovule-wall with the outer part of the nucellus, × 225. Mature seed-coat in t.s., with remains of perisperm, the tannin-cells of endotesta and endotegmen stippled, × 225.

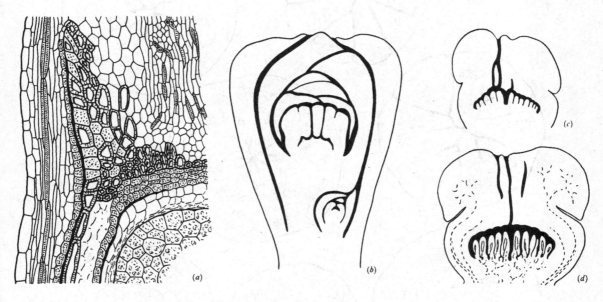

534 *Scyphostegia borneensis*: (*a*) raphe-side of the chalaza in l.s., showing testa, tegmen, perisperm and endosperm, with the tannin-cells of the endotesta and endotegmen hatched, × 120; (*b*) two very young flower-buds in l.s., × 25; (*c*) flower-bud in l.s. to show the centrifugal development of the ovules from the stem-apex, × 25; (*d*) mature ovary in l.s. to show the centri-fugal direction of the ovules, × 10.

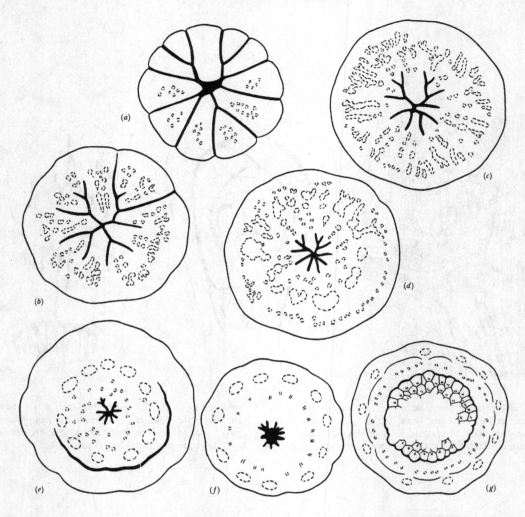

535 *Scyphostegia borneensis.* Ovary in slightly oblique transverse section from the stigmatic surface downwards: (*a*) stigmata derived from the free carpel-primordia; (*b*)–(*e*) the style with stylar canals from the intercarpellary clefts, and the complicated carpellary vasculature, in (*e*) the stigmatic margin separating from the apex of the ovary (*f*); (*g*) the ovary with several ovules, the main cortical v.b. (as the dorsal v.b. of the carpel-primordia), the external limit of the fruit-valves (endocarp), and the small endocarpic v.b.; × 5.

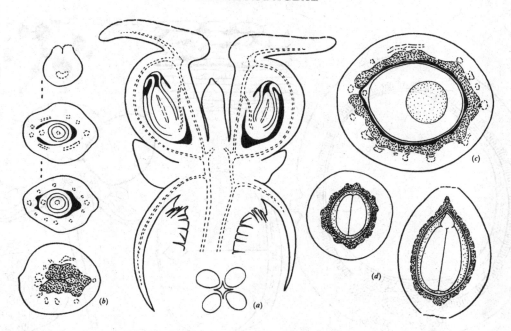

536 *Brucea javanica*. Flower in l.s. just after pollina-tion, showing the vestigial column of the receptacle, the disc and sepals, × 25. Carpel in t.s. (left) from the stigma to the basal part of the ovule, × 25. (*a*) Carpels and receptacular column in t.s., × 6. (*b*) Apex of drupe in t.s., × 5. (*c*) Immature drupe in t.s., with large nucellus, × 10. (*d*) Mature drupe in l.s. and t.s., × 5. Woody endocarp speckled.

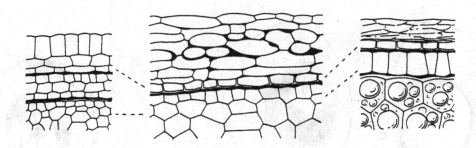

537 *Brucea javanica*. Wall of ovule and the seed-coats of the half-grown and mature seed (right, with endo-sperm), in t.s., × 225.

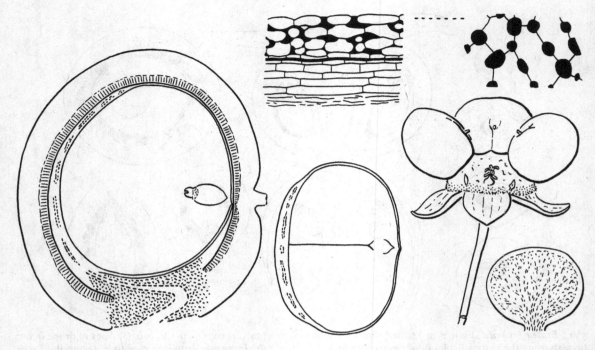

538 *Picrasma javanica*. Fruit with the styles attached to the abortive carpel, × 2. Drupe in l.s. with endocarp and woody hilar tissue, × 8. Seed in t.s., × 5. Chalazal view of seed with its fan of v.b., × 3. Mature seed-coat with aerenchymatous exotesta and remains of the nucellus, × 225.

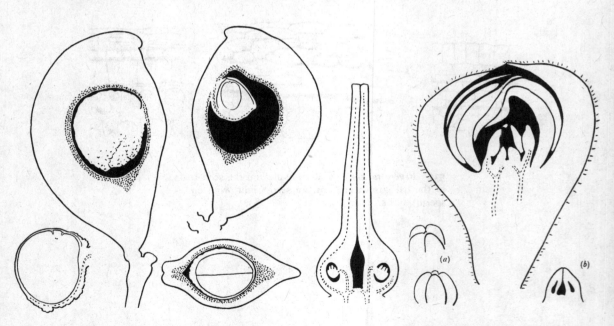

539 *Quassia indica*. Mature and immature fruits in l.s., the woody endocarp speckled, × 1. Mature seed in l.s. and fruit in t.s., × 1. Immature ovary in l.s. and flower-bud in l.s. to show the free carpels and connate style as a subsequent growth, × 25. (a) Very young carpels in side-view, and (b) in l.s., × 25.

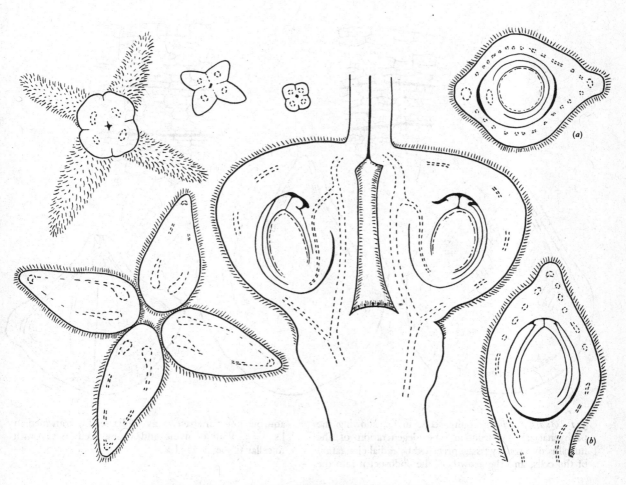

540 *Quassia indica*. Ovary at anthesis in l.s. and (left) in t.s. from the style to the top of the carpels, ×25: (*a*) carpel in t.s., and (*b*) in transmedian l.s., ×25.

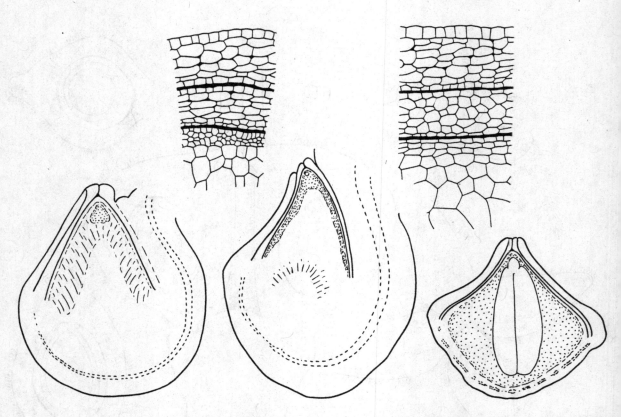

541 *Quassia indica*. Young seeds in l.s. showing the pachychalazal construction, the degeneration of the nucellus into watery tissue preceded by radial elongation of the cells, and the growth of the endosperm into the aqueous nucellar space, × 25. Older seed in transmedian l.s., × 5. Wall of ovule and young seed in t.s. with nucellar tissue, × 225.

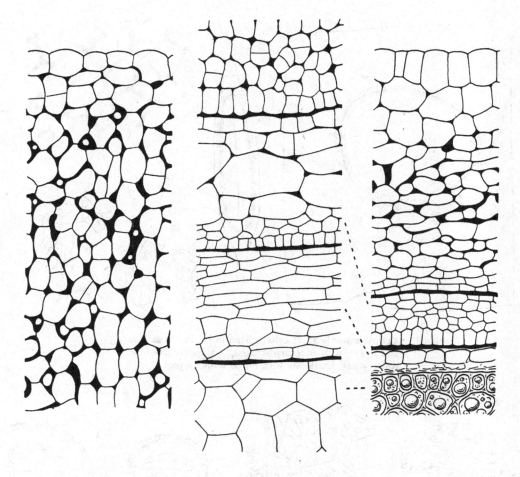

542 *Quassia indica*. Seed-coats of the mature seed with endosperm (right) and the immature seed (left, outer part of testa); centre, inner part of testa with tegmen, nucellus, and endosperm), × 225.

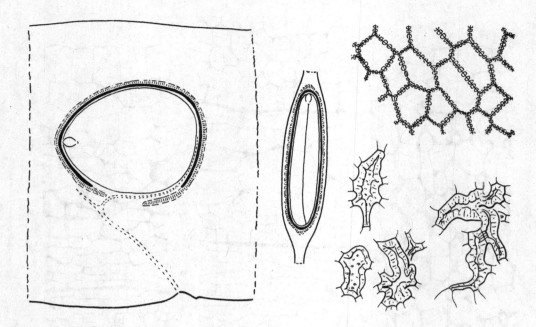

543 *Simarouba sp.* (Ceylon). Seed in l.s. in the samara and in t.s., with endocarp striated, × 2. Sclereids of the testa, × 225. Endotesta with pitted walls, × 500.

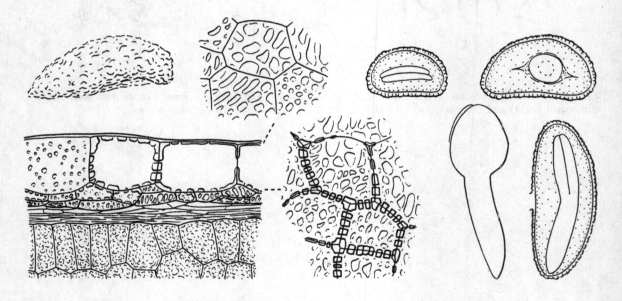

544 *Cestrum aurantiacum.* Seed in side-view, × 8; in section, and the embryo, × 10. Mature seed-coat in t.s. with endosperm, × 225.

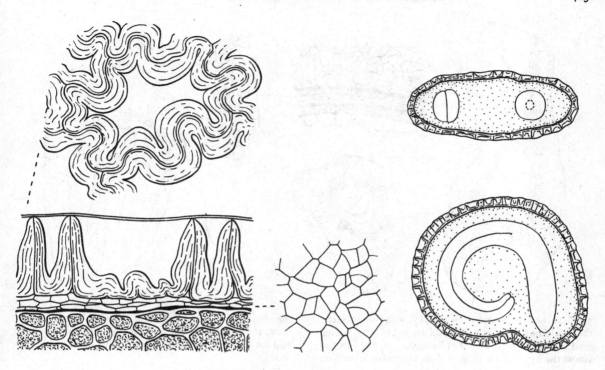

545 *Withania sp.* Seed in l.s. and t.s., × 25. Seed-coat in t.s. with endosperm, with exotestal and endotestal facets, × 225.

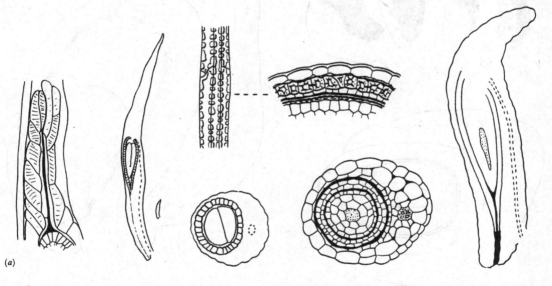

546 *Duabanga taylori.* Seed in l.s., with ovule to scale, × 12. Ovule in l.s., × 115; in t.s., × 225. Seed in t.s., × 50. Seed-coats in t.s. with exotegmic fibres, × 225.

(*a*) Micropyle of the seed of *D. grandiflora*, × 180 (after Venkateswarlu, 1937*a*).

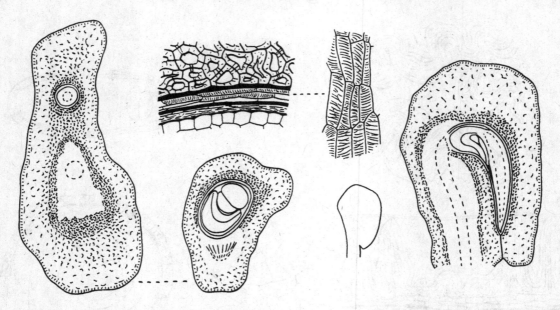

547 *Sonneratia caseolaris.* Seed in l.s. with thin tegmen, thick sclerotic testa and exotesta (striate), × 8. Young seed in outline soon after fertilization, × 8. Seed in t.s. across the hypocotyl and in oblique tangential section at the chalaza, × 12. Seed-coats in t.s. to show the endotesta, the thin tracheidal tegmen, the crushed nucellus and the endosperm, with the epidermis of the embryo, × 225.

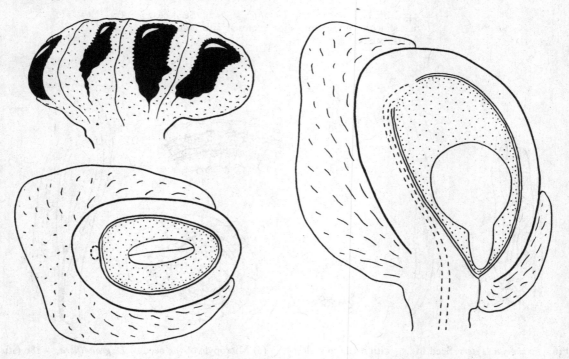

548 *Stachyurus praecox.* Cluster of arillate seeds, × 10.
Seed in l.s. and t.s., × 25.

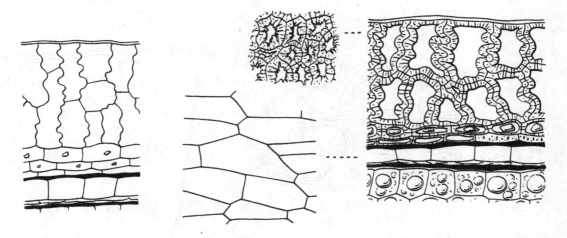

549 *Stachyurus praecox*. Immature and mature
seed-coats, ×225.

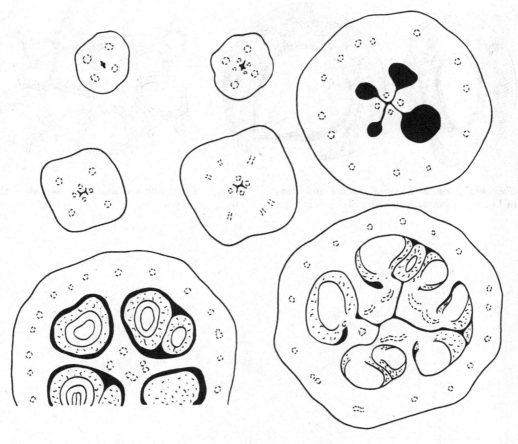

550 *Stachyurus praecox*. Style in t.s. (4 sections), ×25.
Fruit in t.s. from apex to base (3 sections), ×10.

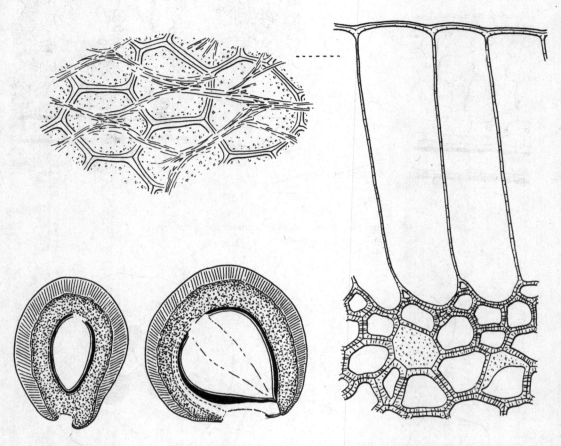

551 *Euscaphis japonica*. Seed in median and trans-median l.s., with exotestal palisade and sclerotic meso-testa, × 8. Palisade and adjacent mesophyll of the testa in t.s., × 225.

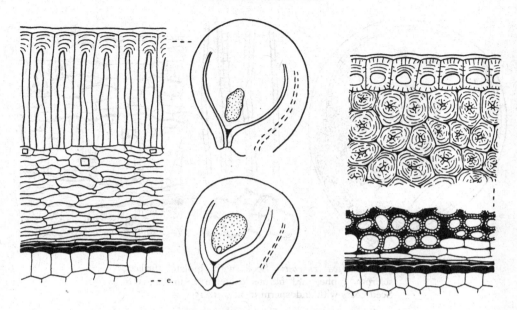

552 *Melianthus major* (left) and *Staphylea pinnata* (right); ovules soon after fertilization, × 25; seed-coats in t.s., × 275 (after Guérin 1901).

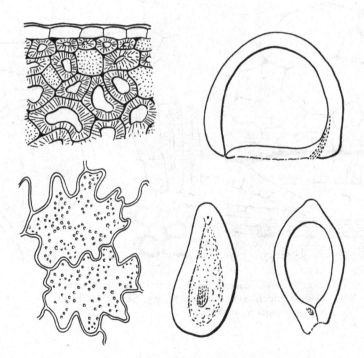

553 *Turpinia grandis*. Seed (empty) in hilar view, and in median and transmedian l.s., × 5; the hard sclerotic testa, × 225; exotestal facets, × 500.

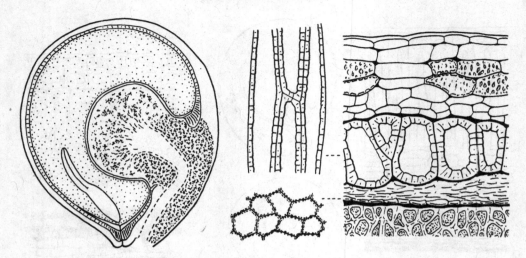

554 *Huertea cubensis*. Seed with bullate chalaza, sclerotic raphe, and fibrous exotegmen, in l.s., ×8. Seed-coats with endosperm in t.s., ×225.

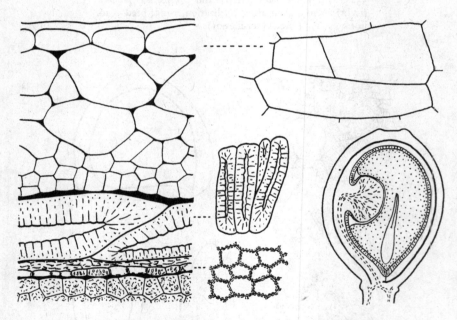

555 *Tapiscia sinensis*. Seed, in the fruit, with bullate chalaza and fibrous exotegmen, ×5. Seed-coats in l.s. with endosperm, ×225.

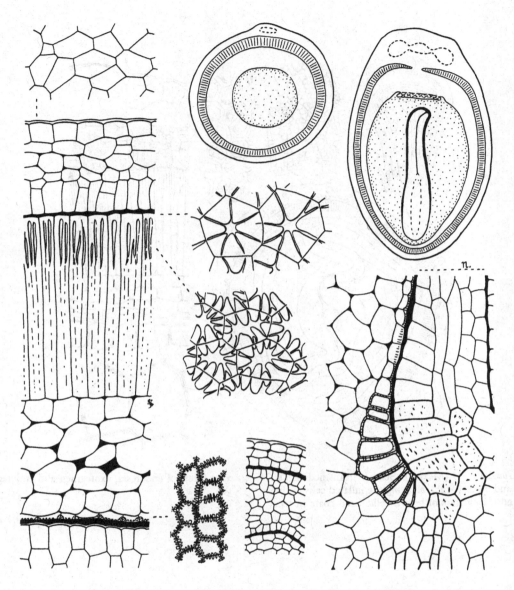

556 *Fremontia californica*. Mature seed in transmedian l.s. and immature seed in t.s., × 10. Wall of ovule and mature seed-coats (left) in t.s., and the outer part of the chalaza in l.s. with nucellus and endotegmen, × 225; cell-details of the exotegmen in t.s., × 500.

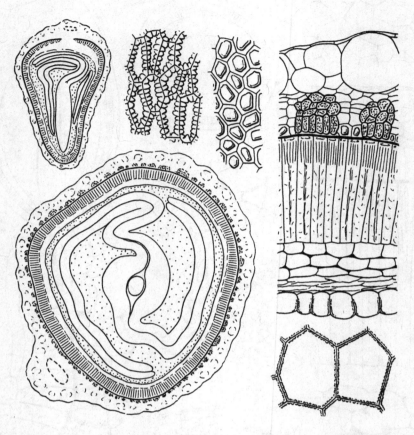

557 *Guazuma sp.* (Mexico). Seed in transmedian l.s. ($\times$ 10) and in t.s. ($\times$ 25), showing the inflated cells of the testa and the clusters of brown cells. Seed-coats in t.s., $\times$ 225; facets of endotesta, exotegmen and endotegmen, $\times$ 500.

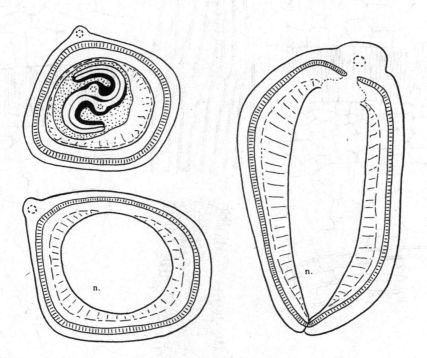

558 *Helicteres sacarolha.* Fully grown but immature seed in l.s. and t.s., and nearly mature seed with embryo and nucellar remains in t.s., ×25.

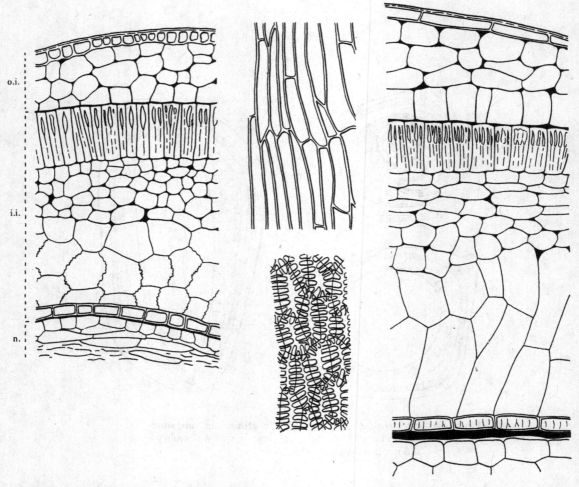

559 *Helicteres sacarolha*. Seed-coats in t.s. of a nearly mature seed (left) and in l.s. of an immature seed (right), ×225; facets of exotesta, ×225; exotegmic palisade-cells in t.s., ×500.

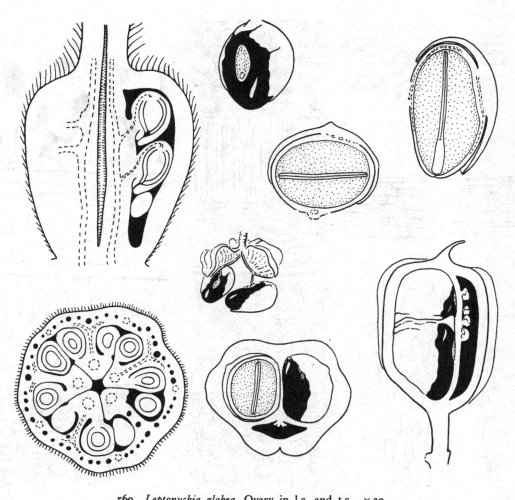

560   *Leptonychia glabra*. Ovary in l.s. and t.s., × 30.
Dehisced fruit, × 1. Ripe fruit in l.s. and t.s., × 2. Seed
in hilar view, × 4; in l.s. and t.s., × 3.

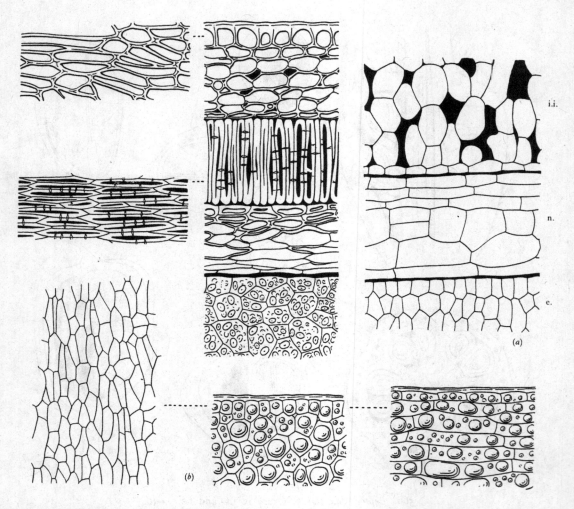

561 *Leptonychia glabra*. Seed-coats with endosperm in t.s., × 225. (*a*) Inner part of the seed-coats of an immature seed in l.s., × 400. (*b*) Outer part of the aril in t.s. and l.s. (right), × 225.

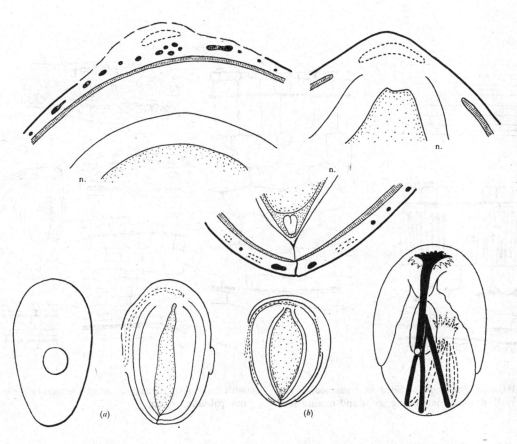

(a)   *(b)*

562 *Leptonychia glabra*. Hilum in t.s., chalaza and micropyle in transmedian l.s., with the exotegmen striated, × 15. (*a*) Immature seeds in antiraphe-view to show the circular inception of the aril, and in l.s., × 15.

(*b*) Immature but fully arillate seed in l.s., × 3. Vascular supply of the testa, with outlines of the aril and hilum, × 5.

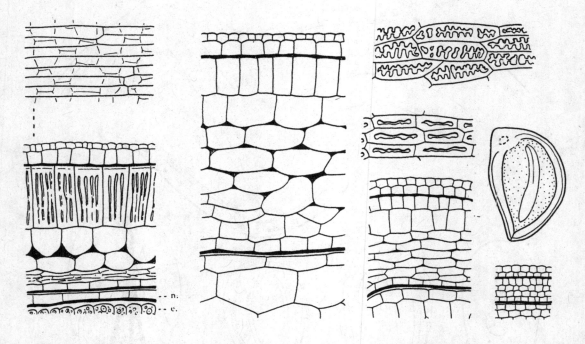

563 *Melochia corchorifolia*. Seed in transmedian l.s., × 10. Wall of ovule, developing seeds and mature seed with endosperm, × 225; facets and section of the exotegmic palisade, × 500.

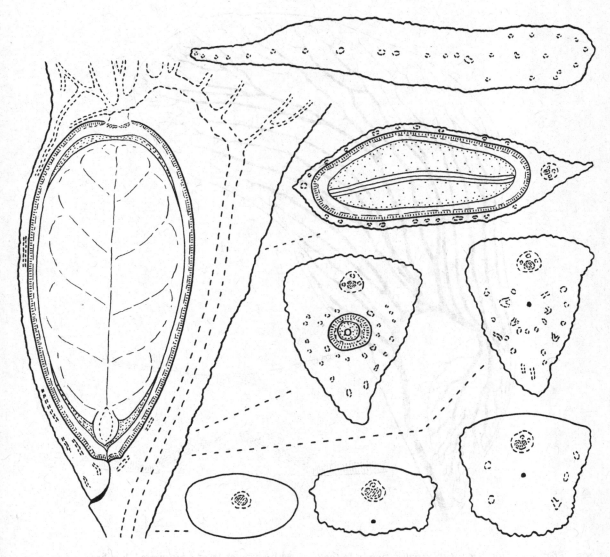

564 *Pterygota alata*. Seed-body in l.s., × 4. Seed in t.s. at 7 levels from the funicle to the base of the wing (uppermost section), × 4. Exotegmen striated.

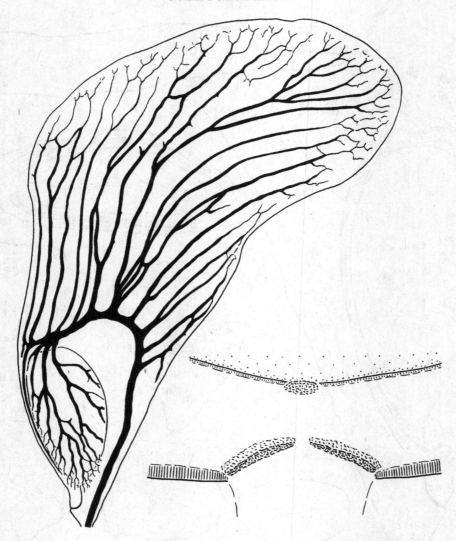

565 *Pterygota alata*. Seed with vascular supply to the wing and to the body (with the v.b. shown only on one side of the body), × 4. Chalaza in l.s. to show the inner hypostase (thin-walled, lignified) and outer hypostase with an outer layer (thin-walled, lignified) and an inner layer (brown walls, not lignified), × 25; exotegmen and endotegmen, striated.

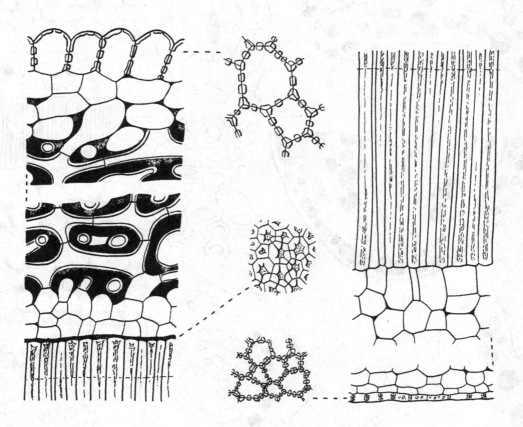

566 *Pterygota alata*. Seed-coats from the seed-body in t.s., testa (left) and tegmen (right), with exotestal and exotegmic facets, × 225; endotegmic facets, × 500.

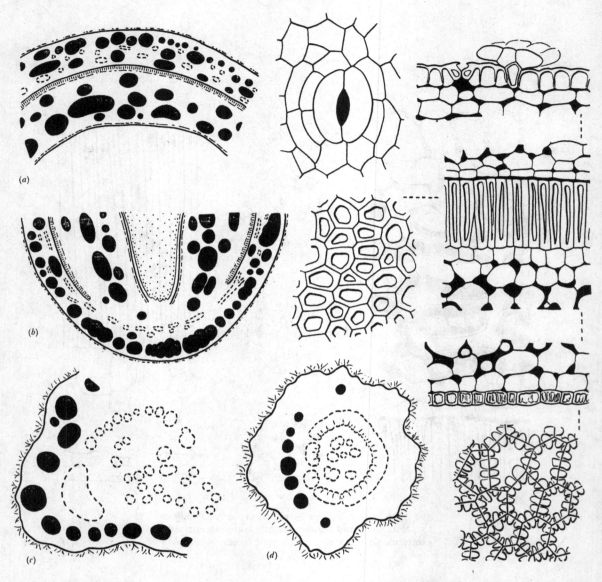

567 *Scaphium macropodum*: (*a*) seed-coat in t.s., the mucilage-sacs in black, × 10; (*b*) transmedian l.s. of the chalaza of the fully grown but immature seed, × 10; (*c*) seed-base in t.s. to show v.b., × 25; (*d*) pedicel of the follicle in t.s., × 25. Seed-coats in t.s. (× 225), with tangential views (× 500).

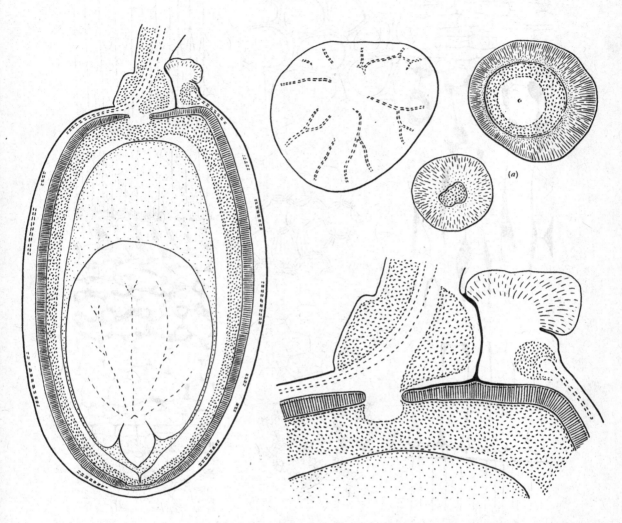

568 *Sterculia foetida*. Seed (nearly mature) in l.s., × 5; exotegmen striated, brown tissue of the funicle and tegmen speckled. Arillostome with false chalaza, × 10. Chalazal end of the seed with terminations of the testal v.b., × 10. (*a*) Chalazal base of the seed and true hypostase in t.s., × 10.

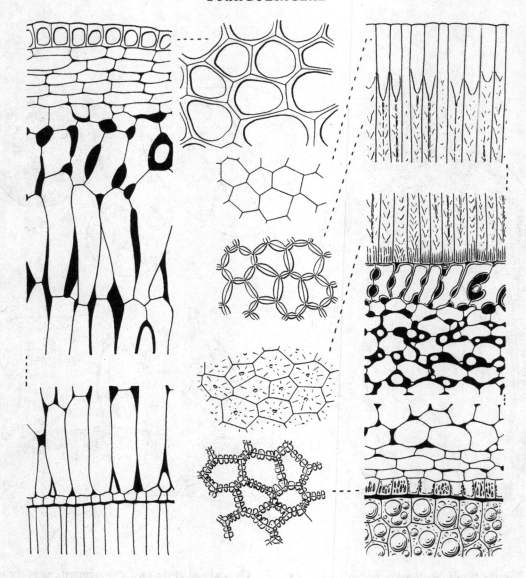

569 *Sterculia foetida*. Testa in t.s. (left), tegmen and endosperm in t.s. (right), × 225. Facets of the exotesta, of the exotegmen (with two sections of the palisade-cells), and of the endotegmen, × 500.

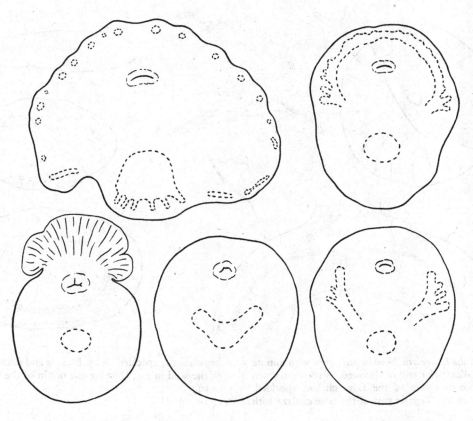

570 *Sterculia foetida*. Funicle with arillostome, and the micropylar end of the seed in t.s. to show the origin of the testal v.b. from an annular bundle round the arillostome, × 10.

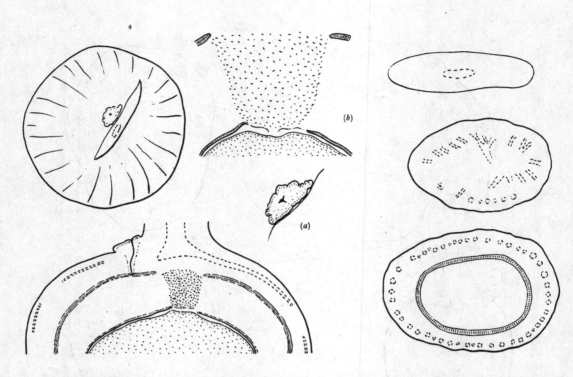

571   *Sterculia lanceolata*. Seed in hilar view with minute aril, × 5. Micropylar end of the seed, × 10; exotegmen striated, brown tissue of the false chalaza speckled. (*a*) Aril, × 10. (*b*) Tegmic part of the false chalaza with brown tissue speckled, × 25. Funicle and micropylar end of the seed in t.s., showing the origin of the testal v.b., × 10.

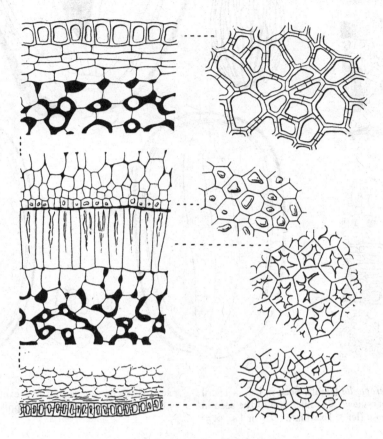

572 *Sterculia lanceolata*. Seed-coats in t.s., ×225;
facets, ×500.

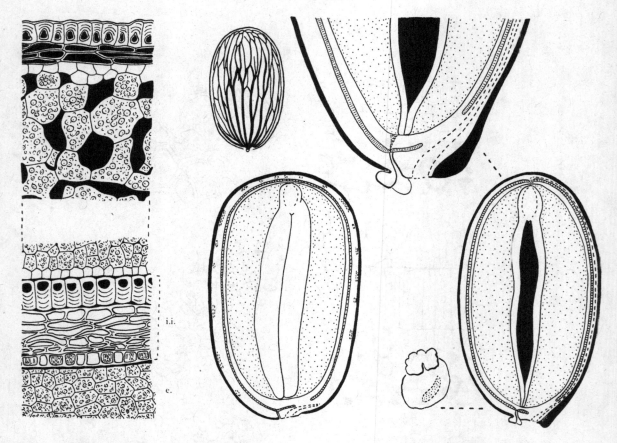

573 *Sterculia macrophylla* (right). Seed in l.s., ×3; aril and hilum in surface-view, ×7; hilar end of the seed in l.s., ×7. *S. rubiginosa* (left). Exarillate seed in l.s., ×5; seed-coats in t.s., with short exotegmic palisade, and endosperm, ×225. Vascular supply to the testa, ×3.

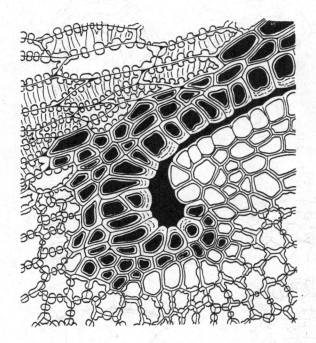 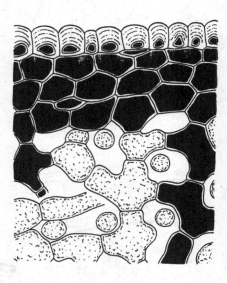

574 *Sterculia macrophylla*. Left, junction of aril and testa with the tracheidal tissue of the hilum, ×225. Right, the outer part of the testa in t.s., ×225; dark contents of the testal cells in black.

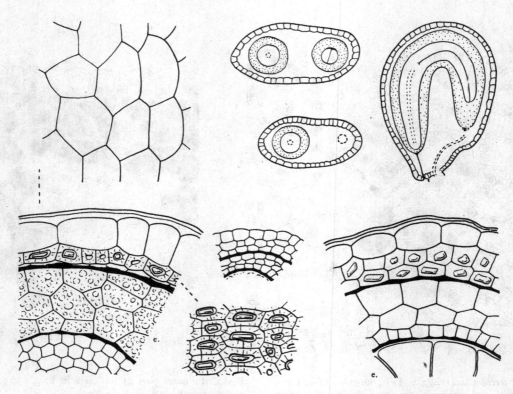

575 *Adinandra impressa*. Ripe seed in l.s. and t.s., ×25. Wall of ovule, of immature seed (right) and mature seed with endosperm, ×225.

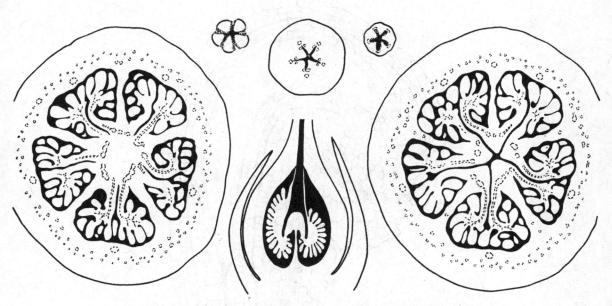

576 *Adinandra impressa*. Ovary in l.s., soon after fertilization, × 5. Stigmata, style and ovary in t.s. at various levels, × 10.

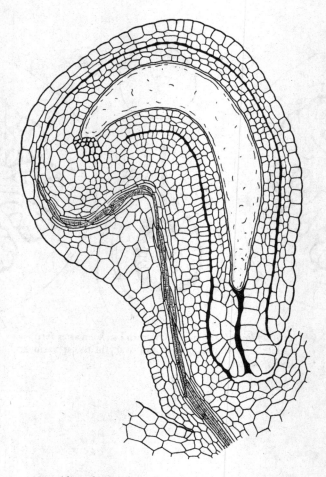

577 *Adinandra impressa*. Ovule in l.s. at anthesis, × 225.

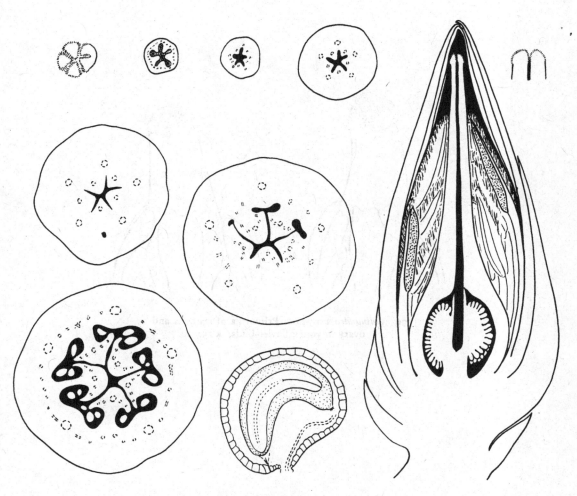

578 *Adinandra verrucosa*. Ripe seed in l.s., ×25. Flower-bud in l.s. ×5. Stigmata, style and ovary in t.s. at various levels, ×10.

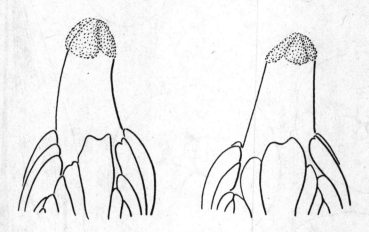

579   *Adinandra verrucosa*. Primordia of stamens and ovary in young flower-buds, × 25.

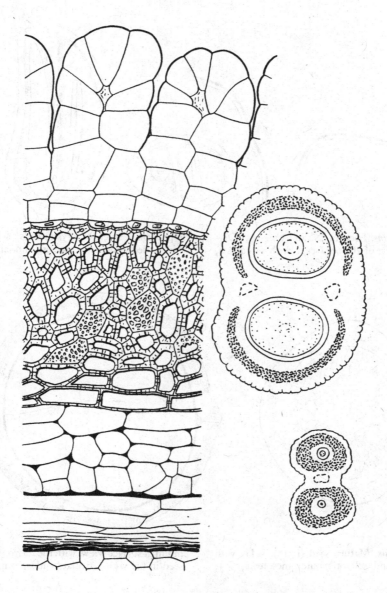

580 *Anneslea crassipes*. Full-sized but immature seed in t.s. across the middle and across the micropyle, × 12.

Seed-coat in t.s. with nucellar cuticle and the outer layer of the endosperm, × 225.

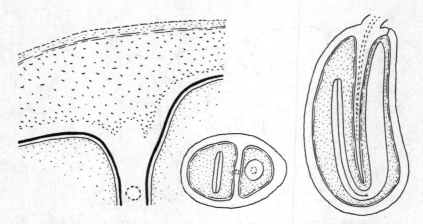

581 *Anneslea fragrans.* Seed in l.s. and t.s., × 5. Diagram of the structure of the seed-coat with sarcotesta, hypodermal layer of crystal-cells, sclerotic mesotesta, thin-walled endotesta, tegmen, and endosperm, × 5.

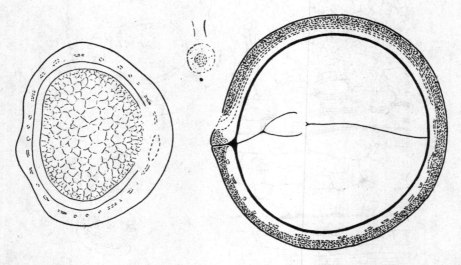

582 *Camellia sinensis.* Mature seed (right) in l.s. with hilum, micropyle, and sclerotic outer mesotesta, × 5; hilum in surface-view, with the micropyle, × 5. Immature seed in t.s. with reticulate endosperm, × 10.

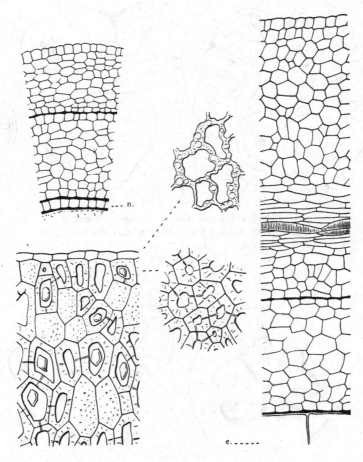

583 *Camellia sinensis.* Wall of ovule with thin nucellus, wall of half-grown seed with large-celled endosperm (right), and the outer part of the testa of the mature seed, in t.s., × 225.

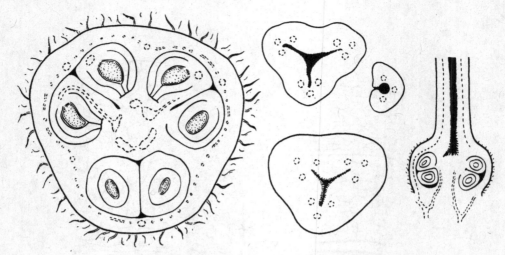

584 *Camellia sinensis*. Ovary of open flower in l.s.
(right, × 10) and in t.s. at various levels from the stigma
and style (× 25).

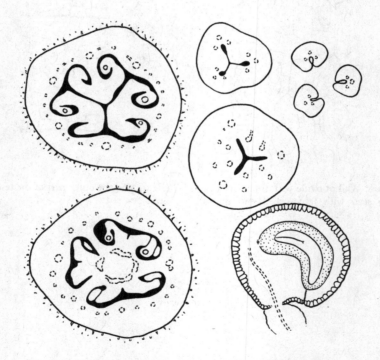

585 *Eurya trichocarpa*. Ripe seed in l.s., and the ovary
of the open flower in t.s. at various levels from the
stigmata, × 25.

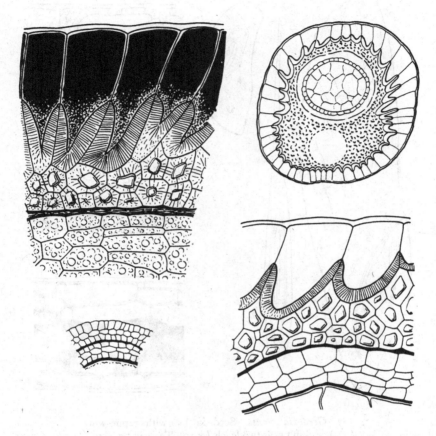

586 *Eurya trichocarpa*. Ovule-wall in t.s., immature seed-coats in t.s., and mature seed-coats with thick-walled cells and endosperm, × 225. Young seed in t.s. to show the large-celled exotesta, the sclerotic mesotesta, the endotegmen and the large-celled endosperm, × 50.

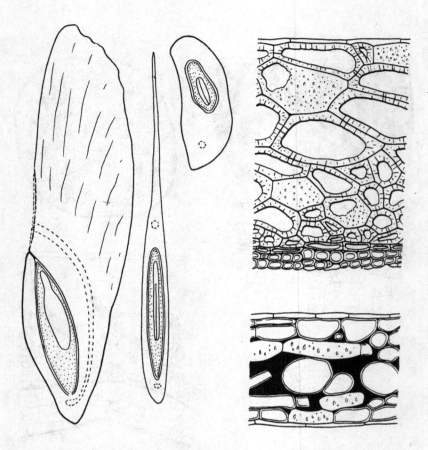

587 *Gordonia obtusa*. Seed in l.s. with raphe-wing
(×5), and in t.s. at two levels (×10). Testa in t.s. across
the body of the seed (above) and across the wing, ×225.

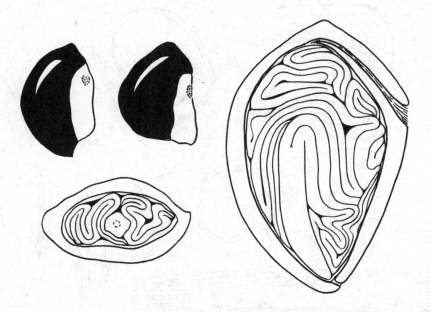

588 *Pyrenaria acuminata.* Mature seed in side-view,
×2; in l.s. and t.s. with sclerotic testa, ×5.

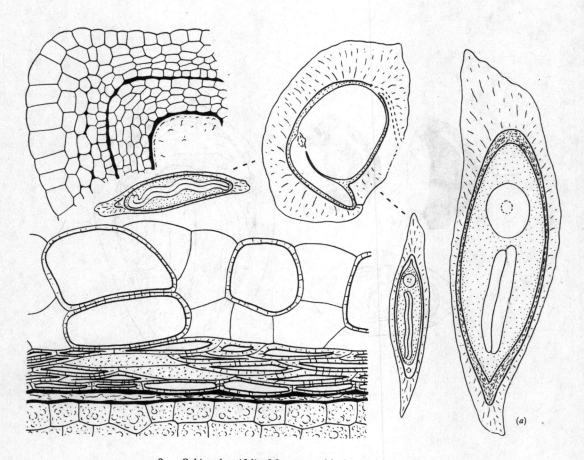

589   *Schima brevifolia.* Mature seed in l.s. and t.s., × 8:
(a) seed in t.s. across the radicle, × 25. Wall of ovule and
of seed with endosperm, × 225.

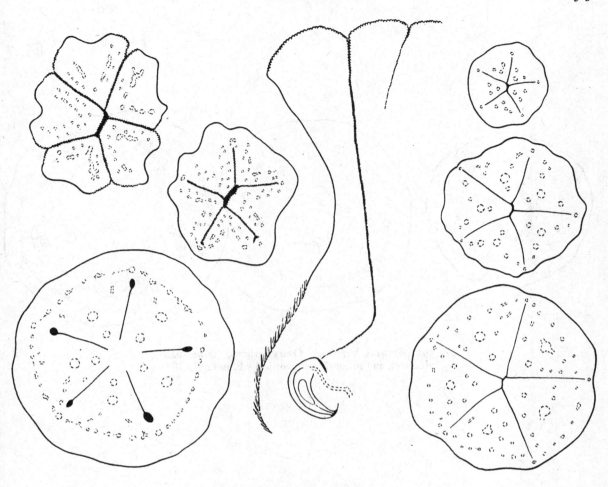

590 *Schima brevifolia*. Part of the ovary of a mature flower-bud in l.s., ×25. Stigmata, style and top of the ovary in t.s., ×18.

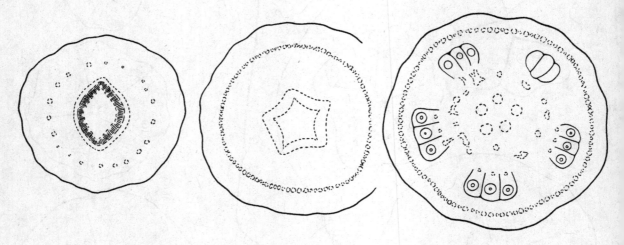

591 *Schima brevifolia*. Ovary (right), flower-base
(centre), and peduncle with cortical v.b. in t.s., ×18.

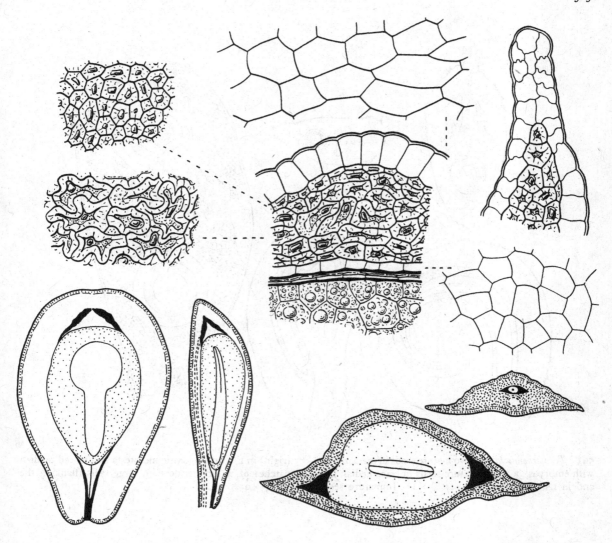

592 *Stewartia koreana*. Seeds in median and trans-median l.s. (× 10) and in t.s. across the cotyledons and the micropyle (× 18). Seed-coats and endosperm in t.s., and wing of seed in t.s., × 225; facets of the exotesta and endotegmen, with tangential views of the mesotesta, × 225.

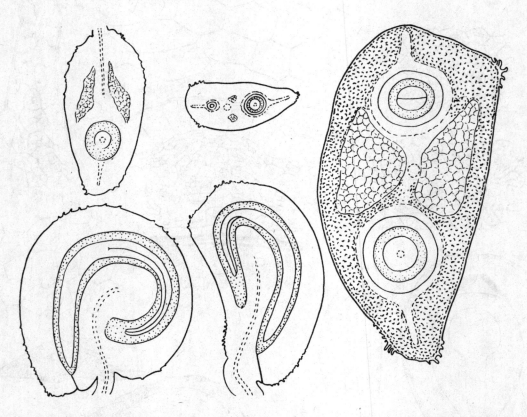

593 *Ternstroemia lowii.* Two mature seeds in median l.s., with embryos, × 10. Immature seeds in transmedian l.s. and in t.s. across the hypostase, × 10. Immature seed (right) in t.s. with sclerotic mesotesta and paired median patches of watery tissue with large cells (flanking the v.b.), × 25.

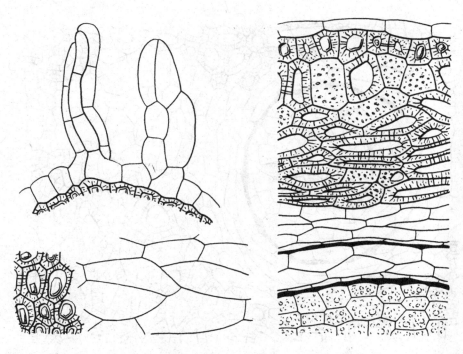

594 *Ternstroemia lowii*. Wall of immature seed in t.s., with sclerotic mesotesta, tegmen and endosperm; marginal papillae of the seed; exotesta and o.h. (testa) in surface-view, × 225.

595 *Ternstroemia lowii.* Hypostase in t.s., with tegmen (i.e. with thick inner walls) and testa (with sclerotic outer mesotesta), × 225.

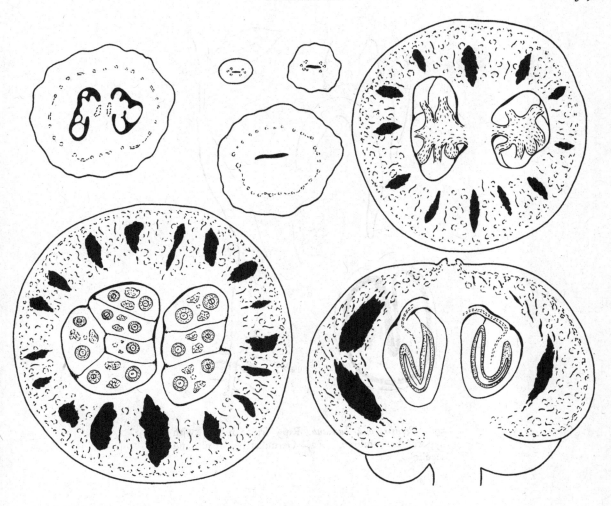

596 *Ternstroemia lowii.* Nearly mature fruit in l.s., and
in t.s. at the level of the placentas and below, × 5. Ovary
and style in t.s., × 10.

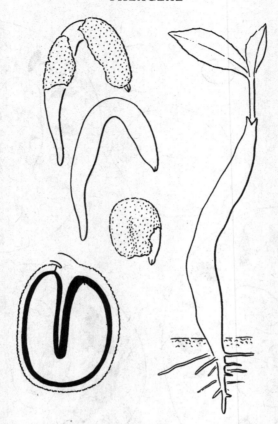

597 *Ternstroemia bancana*. Ripe fruit in l.s., with hypocotylar embryo, × 2. Germinating seed and seedlings, × 1.

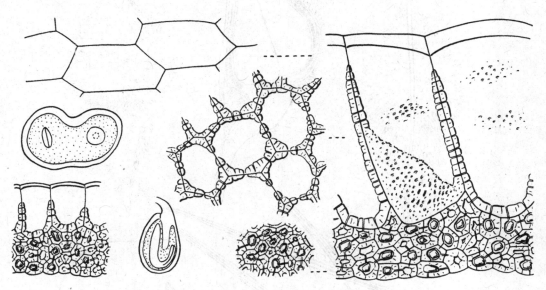

598   *Visnea mocanera*. Seed in l.s. (× 5) and t.s. (× 10).
Testa in t.s. at the thick part (right) and thin part (left),
× 225.

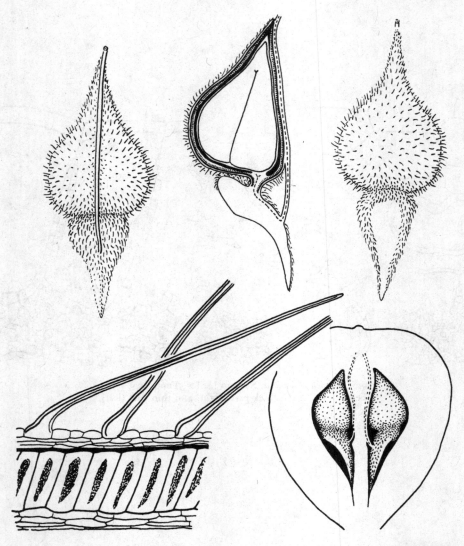

599 *Aquilaria malaccensis.* Seed in raphe-view, l.s. (with tegmen hatched), and antiraphe-view, × 5. Testa and tegmen in t.s., × 225. Fruit in section before dehiscence, × 2.

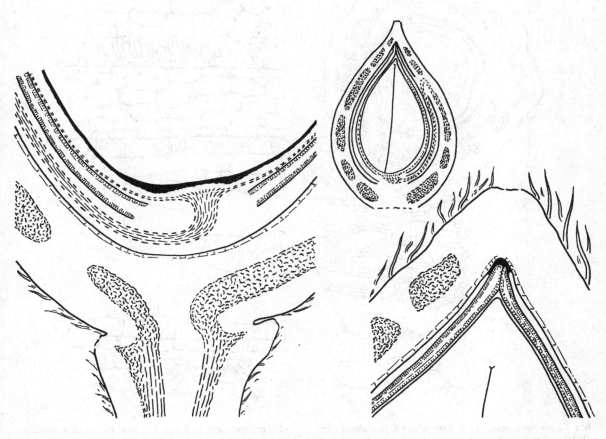

600 *Linostoma pauciflorum*. Drupe in l.s., ×3. Chalaza with nucellar v.b., and micropyle in l.s., ×25. Sclerotic tissue speckled in the pericarp; palisades of o.e. and i.e. (tegmen) and of the fibrous endocarp striated in the enlargements.

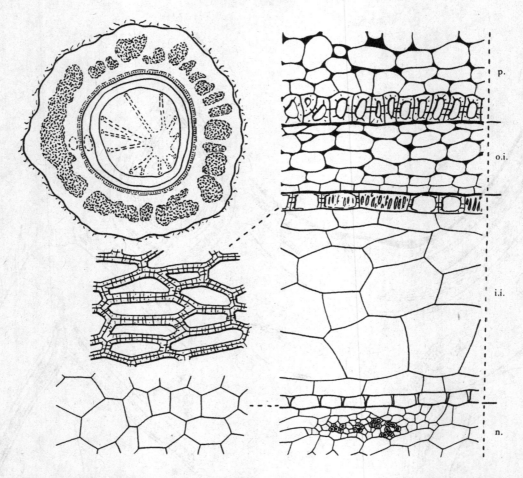

601 *Linostoma pauciflorum*. Drupe in t.s. just above the chalaza with nucellar v.b.; sclerotic tissue of pericarp speckled, endocarp striated. Seed-coats in t.s. with adjacent pericarp (p.) and nucellus (n.) with a nucellar v.b., × 225.

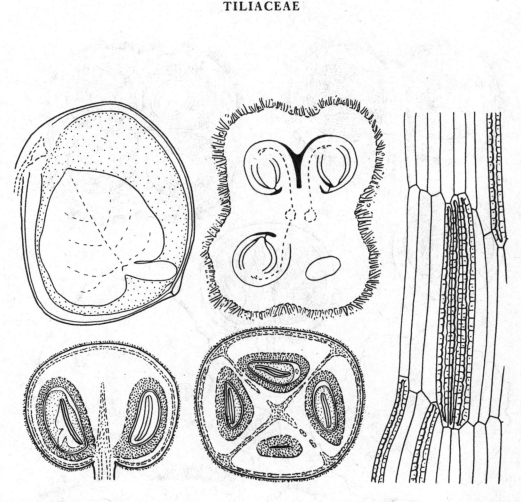

602 *Grewia acuminata*. Seed in l.s., ×8. Ovary in t.s., ×50. Fruit in l.s. and t.s. with the bony pyrenes stippled, ×2. Exotesta in surface-view with thin-walled and sclerotic cells, ×225.

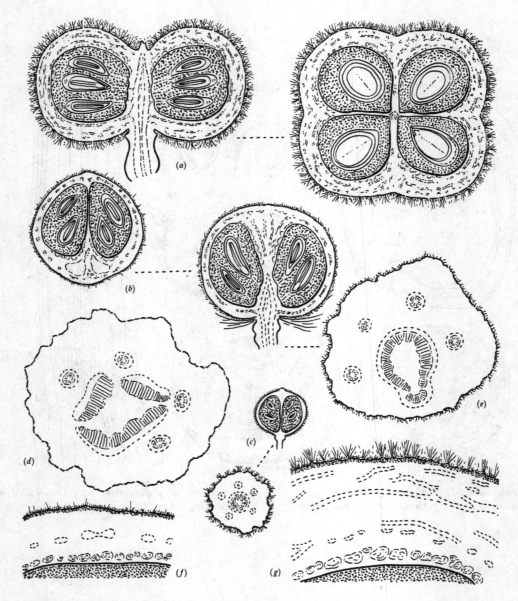

603 *Grewia.* Fruits of (*a*) *G. glandulosa*, (*b*) *G. orientalis* and (*c*) *G. polygama* in t.s. and l.s., showing the bony pyrenes, ×2. Fruiting peduncles, in t.s., of (*d*) *G. glandulosa*, ×18; (*e*) *G. orientalis*, ×25. Pericarp of (*f*) *G. orientalis* and (*g*) *G. glandulosa* in t.s. to show the sclerotic exocarp, the v.b. and the slime-canals external to the pyrenes (stippled), ×10.

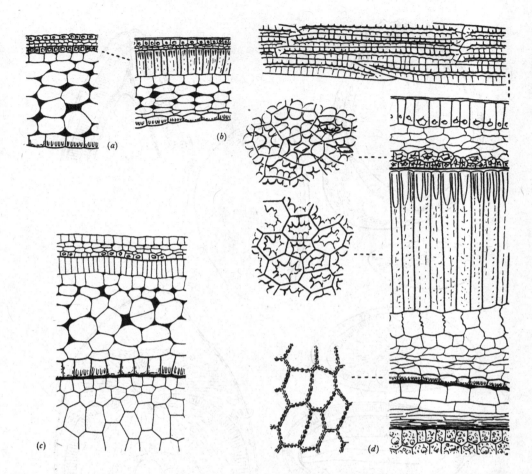

604 *Grewia.* Fully grown seed-coats in t.s., of (*a*) *G. polygama* with very short cells in the exotegmen, (*b*) *G. ?holstii*, (*c*) *G. orientalis* (immature) and (*d*) *G. glandulosa* (fully mature), with endosperm and trace of nucellus, × 225. Surface-view of i.h. (testa, × 225), and facets of the exotegmen and endotegmen, × 500.

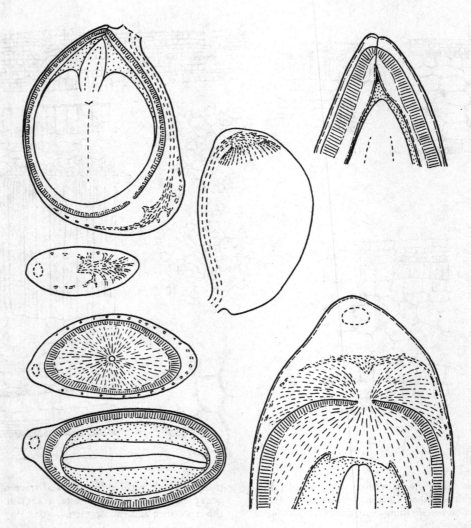

605 *Grewia glandulosa*. Seed in median l.s., × 8; in t.s. at the chalaza, heteropyle, and across the middle, × 10. Micropylar and chalazal ends of the seed in transmedian l.s. with the rows of brown cells in the chalazal tissue shown as broken lines, × 25. Diagram of v.b. and the chalazal umbrella.

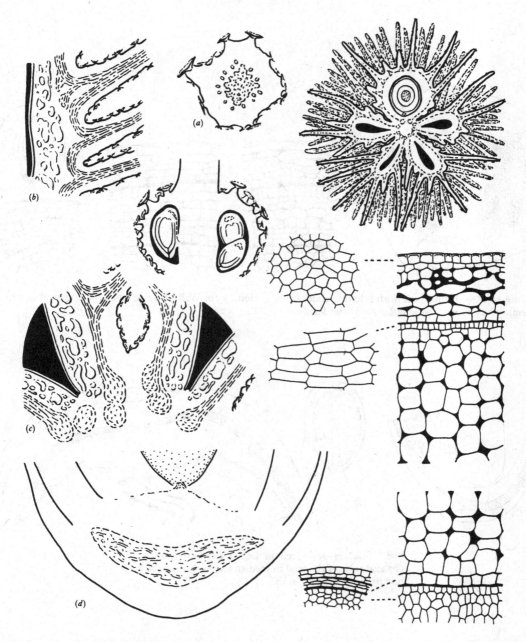

606 *Jarandersonia*. Fruit in t.s., ×2. Ovary in l.s., ×25. (*a*) Fruit-spine in t.s. with mucilage-canals external to the fibrovascular core, ×25. (*b*) Bases of fruit-spines in l.s., ×10. (*c*) Septum between two sterile loculi of the fruit in t.s. with fibrovascular bundles and mucilage-canals, ×10. (*d*) Chalaza in transmedian l.s., ×25. Wall of ovule and seed-coats of the immature seed with nucellar tissue, ×225.

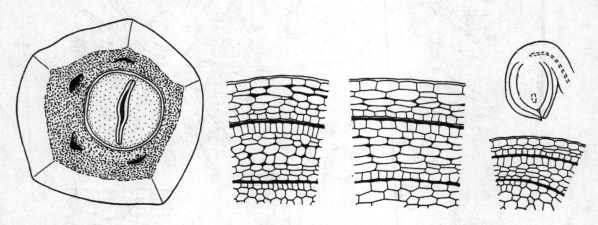

607 *Tilia europea*. Fruit in t.s. with sclerotic endocarp and 4 empty loculi, × 5. Ovule in l.s. soon after fertiliza-tion, × 10. Wall of ovule and of young seed in t.s. and l.s., × 225.

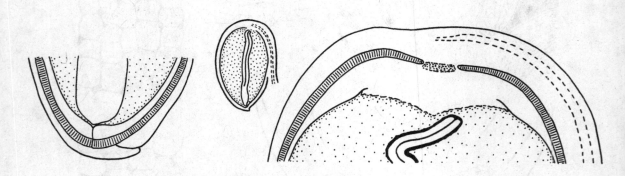

608 *Tilia europea*. Seed in l.s., × 4. Micropylar and chalazal ends of the seed in median l.s., with the exotegmic palisade striated, × 18.

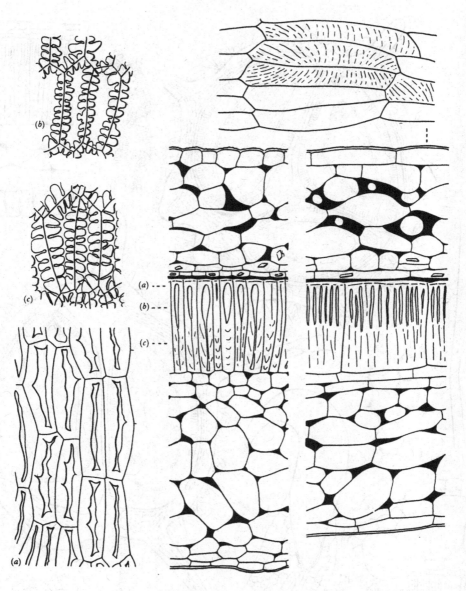

609 *Tilia europea*. Seed-coats in t.s. and l.s. (right),
× 225. Palisade-cells in t.s. at levels (*a*), (*b*) and (*c*), × 400.

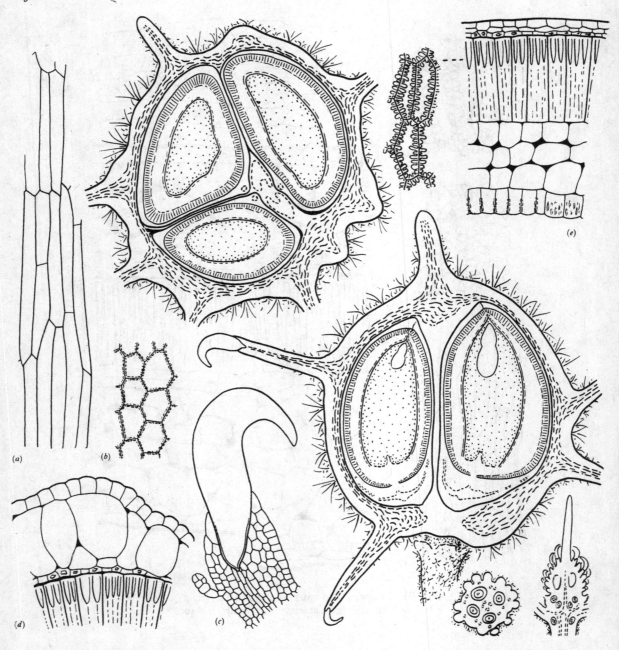

610 *Triumfetta bartramia*. Fruit in l.s. and t.s., fully grown but with immature embryos, showing the fibrous endocarp, and the ovary in l.s. with incipient spines and large slime-canals in the receptacle, ×18; ovary in t.s., ×25: (*a*) exotesta; (*b*) endotegmen; (*c*) young spine of ovary with terminal hooked hair; (*d*) seed-coats in t.s. at the free angle of the seed with enlarged mesotesta; (*e*) seed-coats in t.s. on the side of the seed; ×225.

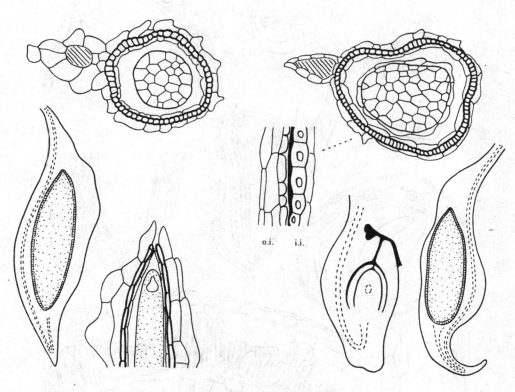

611   *Tetracentron* (left) and *Trochodendron*; seeds cleared
and in section; (after Bailey and Nast 1945, mag.).

o.i.   i.i.

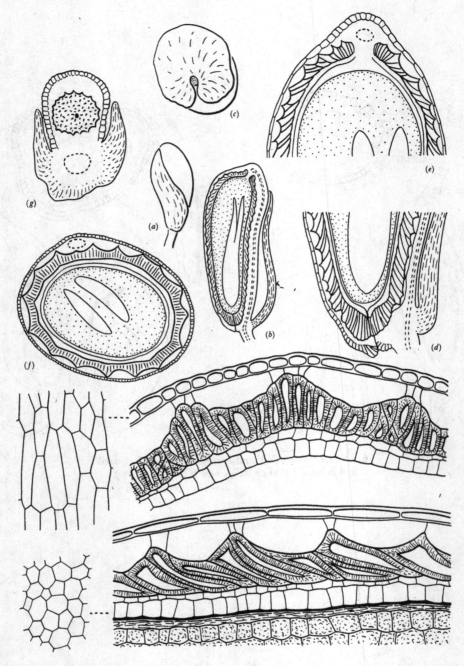

612 *Turnera ulmifolia*: (*a*) seed with aril and funicle, × 5; (*b*) seed in l.s. with aril, × 17; (*c*) base of seed with aril, × 25; (*d*) micropylar end of the seed in l.s., × 50; (*e*) chalazal end of the seed in transmedian l.s., × 50; (*f*) seed in t.s., × 50; (*g*) junction of aril with funicle in t.s. and micropylar end of the seed, × 50. Seed-coats in t.s. (upper) and l.s. (lower), showing the large cells of the endotesta in the depressions of the sclerotic exotegmen and, in the lower figure, the remains of the nucellus and outer layer of the endosperm, × 225.

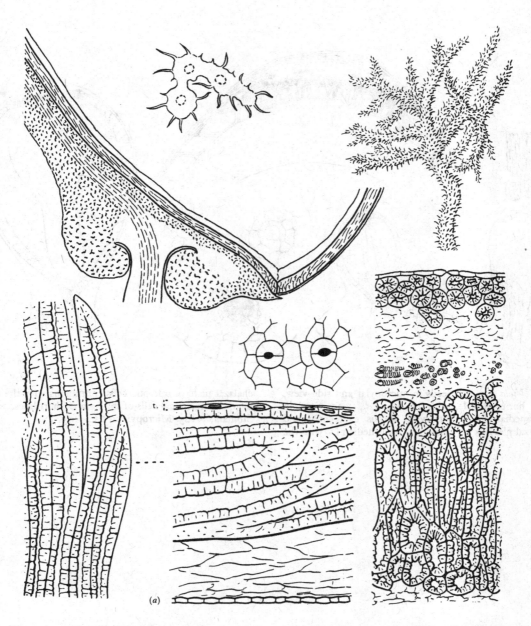

(a)

613 *Rinorea anguifera*. Fruit-process, × 5; in t.s. with 3 v.b., × 25. Funicle and micropyle in l.s., showing the sclerotic tissue of the testa in the hilar region, the arillar swelling, and the fibrous exotegmen, × 25. (*a*) Seed-coats in l.s. from the side of the seed, with thin testa (t.) and fibrous tegmen, × 225. Chalazal stomata, × 225. Chalazal plate in t.s. (lower right), × 225.

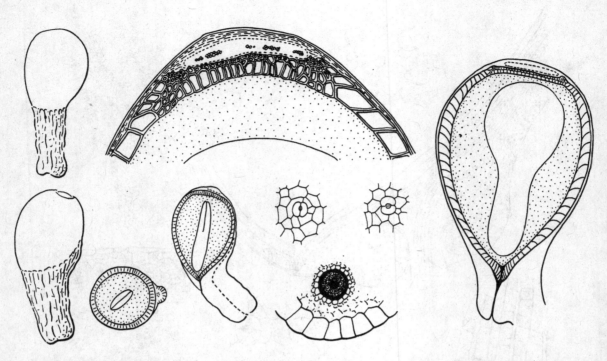

614 *Viola odorata*. Seeds in micropylar and side-view, with horny funicle, with l.s. and t.s., × 10. Seed in transmedian l.s. (right) with fibrous exotegmen and chalazal plate, × 25. Chalaza in transmedian l.s. with the chalazal plate of sclerotic cells continuous with the tegmen, the tracheidal cells of the testa as short lines, × 75. Stomata, × 225. Micropyle in t.s., × 75.

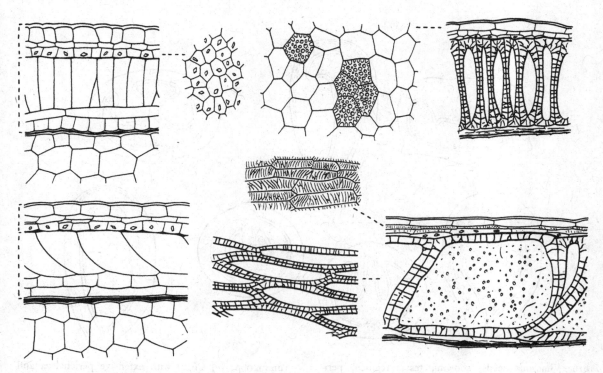

615 *Viola odorata*. Immature seed-coats (left) in t.s. and l.s., with endotestal crystal-cells and nucellar tissue, × 225. Mature seed-coats (right) in t.s. and l.s., with the exotesta, o.h. (testa) and exotegmen in surface-view, × 225.

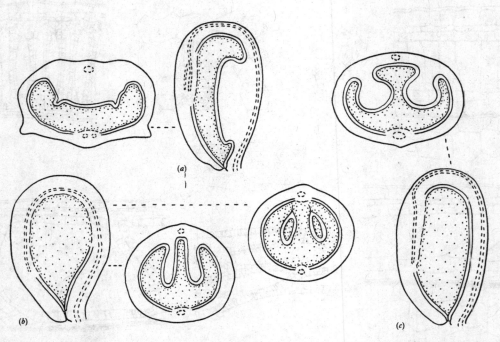

616 Vitaceous seeds, showing testa, tegmen, perichalaza, and endosperm, mag. (after Periasamy 1962a):
(a) *Cayratia* with shortly extended chalaza and slight
rumination; (b) *Cissus* with extensive perichalaza and
deep rumination; (c) *Tetrastigma* with intermediate perichalaza and rumination.

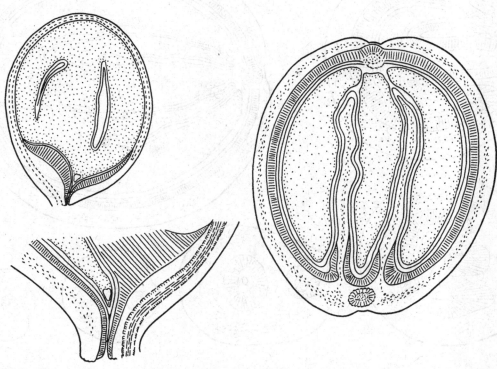

617 *Cissus quadrangularis*. Seed, fully grown but immature, in median l.s. (left), ×10; in transmedian l.s. (right), ×18. Micropyle in median l.s., ×25. Woody endotesta hatched. Raphid-cells shown as short dashes in the sarcotesta.

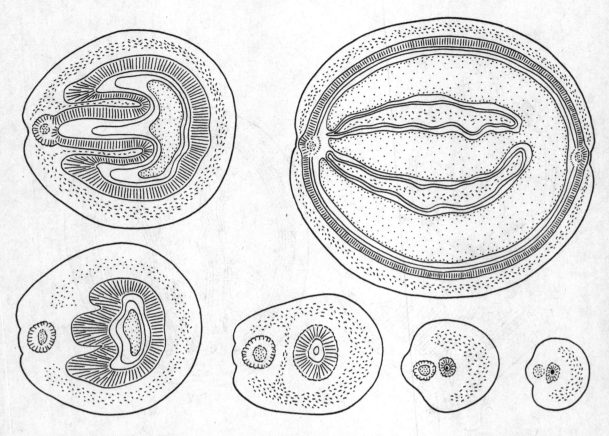

618 *Cissus quadrangularis*. Seed, fully grown but immature, in t.s. from the micropyle (bottom right) to the centre, × 18. Woody endotesta hatched. Raphid-cells as short dashes in the sarcotesta. Raphe v.b. on the left in the sections.

619 *Cissus quadrangularis*. Seed, fully grown but immature, in t.s. from the central part to the chalazal end (upper left), × 18. Woody endotesta hatched. Raphid-cells as short dashes in the sarcotesta. Raphe v.b. on the left in the sections.

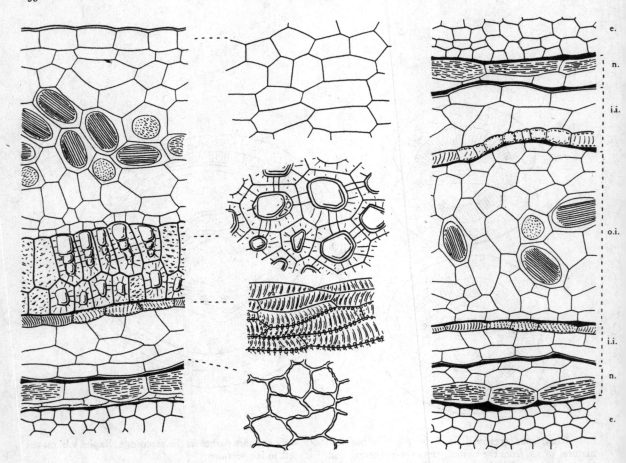

620 *Cissus quadrangularis.* Seed-coat in t.s. (left) and a rumination in t.s. (right), × 225. Details of testa and tegmen in surface-view, × 500.

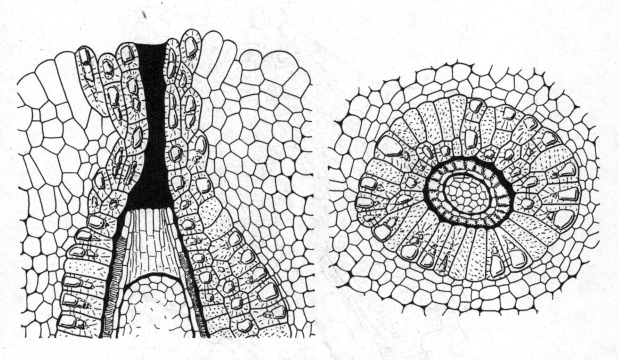

621   *Cissus quadrangularis*. Micropyle of the seed in l.s.
and t.s., × 225.

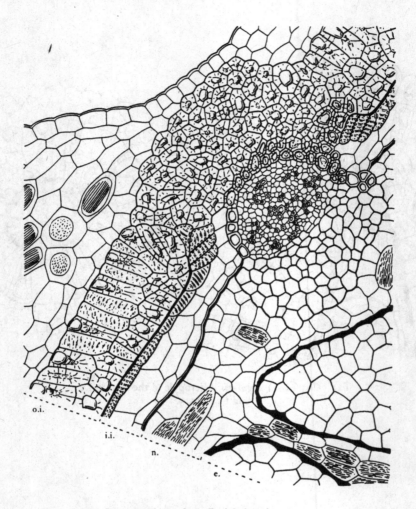

622 *Cissus quadrangularis.* Perichalaza in t.s., ×225.

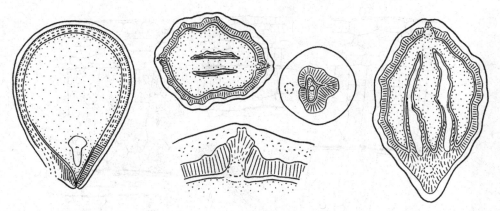

623 *Cissus sp.* (Ceylon). Seed in median (left) and
transmedian (right) l.s. and in t.s. across the centre and
at the micropyle, × 8. Perichalaza in t.s., with raphid-
cells in the sarcotesta, × 25. Woody endotesta striated.

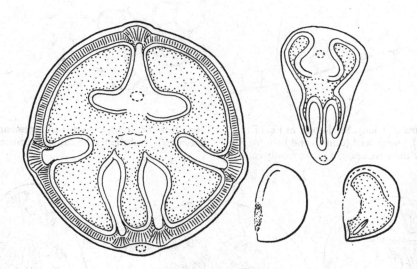

624 *Leea sambucina* (Ceylon). Mature seed in t.s. and
abortive seed (upper right) in t.s., × 18. Mature seed in
side-view and l.s., × 5. Woody endotesta striated.

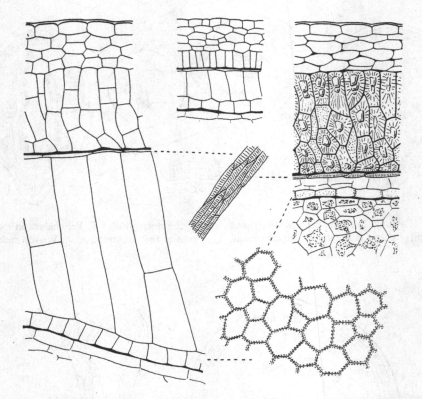

625 *Leea sambucina* (Ceylon). Seed-coats in t.s. of the mature seed (right), young seed (centre) and half-grown seed (left), to show the development of the woody endo-testa from i.e. (o.i.), the great enlargement of the tegmic mesophyll, and the tracheidal exotegmic cells, × 225.

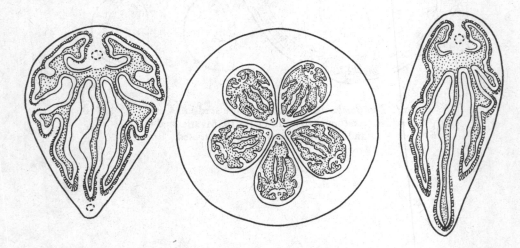

626 *Leea sp.* (RSS 6037). Ripe fruit in t.s., × 2. Seed, fully grown but immature, in t.s. (left) and transmedian l.s. (right), with the nucellus stippled, × 10. Endotesta striated.

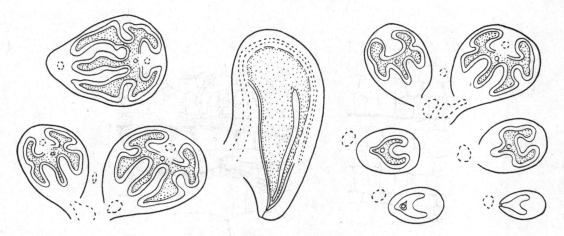

627 *Leea sp.* (RSS 6037). Seed, half-grown, in l.s. and
t.s. from the micropyle (right) to the centre (left), × 10;
nucellus stippled.

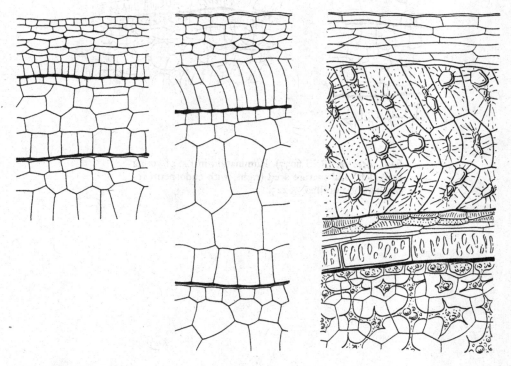

628 *Leea sp.* (RSS 6037). Seed-coats in t.s. of young
seed (left), fully grown but immature seed (centre) and
mature seed (right), with endosperm replacing the
nucellus, × 225.

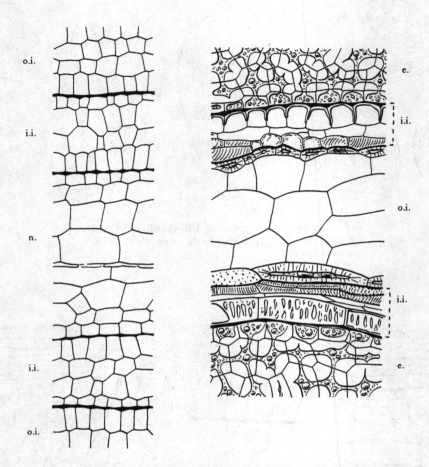

629 *Leea sp.* (RSS 6037). Rumination in t.s. of young
seed (left) and mature seed (right) with endosperm replacing the nucellus, × 225.

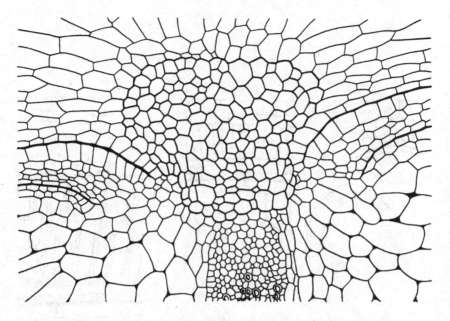

630 *Leea sp.* (RSS 6037). Perichalaza of a young seed in t.s., the nucellar tissue uppermost, × 225.

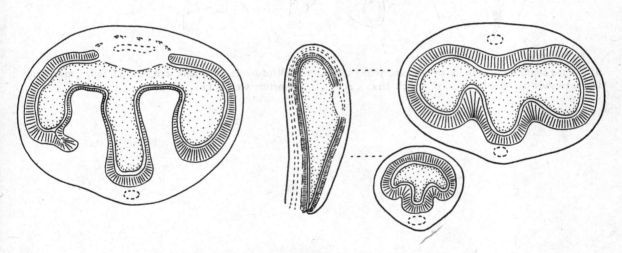

631 *Vitis vinifera.* Full-grown seed in l.s. (centre), × 10; in t.s. at the chalaza (left), and above and below the chalaza (right), × 18.

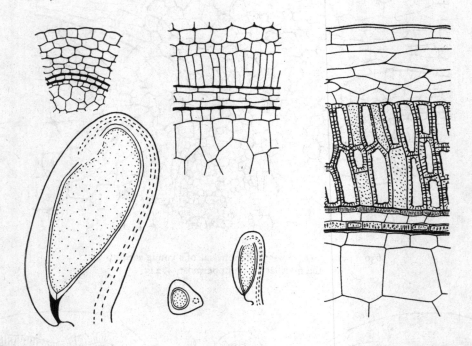

632   *Vitis vinifera.* Ovule in t.s. and l.s., and young seed in l.s. to show the displacement of the chalaza, ×25.

Wall of ovule, of young seed, and of fully grown but immature seed in t.s., ×225.

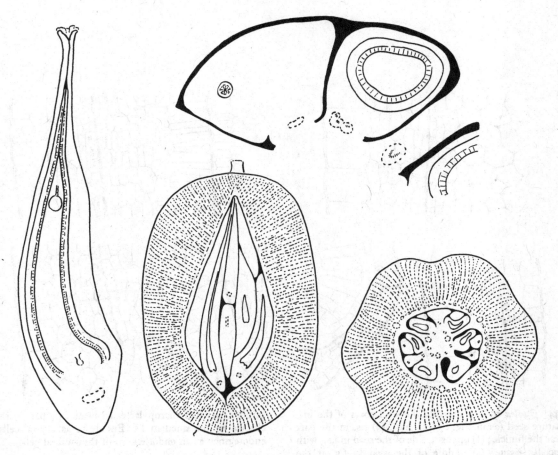

633 *Qualea sp.* Immature fruits in t.s. and tangential l.s. with the exocarp exfoliated and the mesocarp traversed by fine radiating v.b., × 5. Immature seed of this fruit in transmedian l.s. and in t.s. (in the fruit-loculi) with the endotesta striated, × 25.

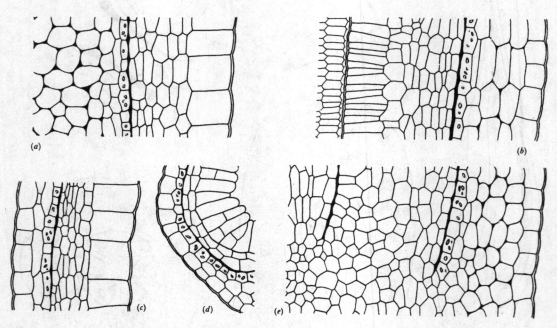

634 *Qualea sp.* Structure of the seed-coats of the immature seed (as in Fig. 633), ×225: (*a*) t.s. in the part near the funicle; (*b*) opposite side of the seed in l.s., with nucellar tissue; (*c*) middle of the seed in t.s. at the thinner part; (*d*) micropylar canal in t.s.; (*e*) part of the chalaza in transmedian l.s. Endotesta as crystal-cells; endotegmen as an endothelium of thin-walled cells.

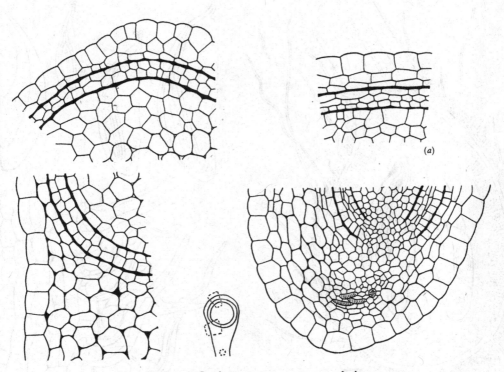

635 *Qualea sp.* Ovular structure, ×225; ovule in t.s.,
showing the enlarged parts with nucellar tissue; (*a*)
ovule-wall in l.s. Chalaza in transmedian l.s.

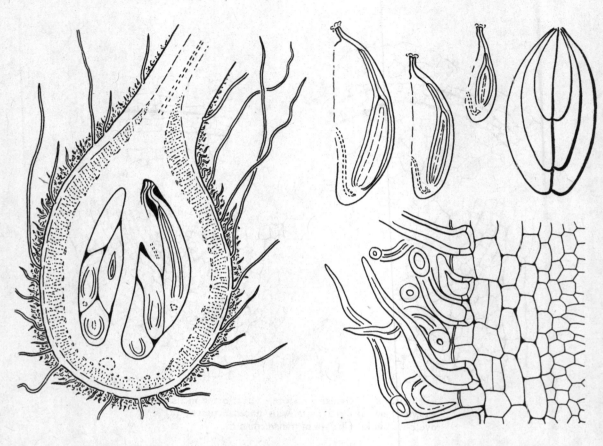

636 *Qualea sp.* Ovary in tangential l.s. to show the long ovules, the incipient v.b. of the mesocarp, and the exocarp before exfoliation, × 25. Ovules on the placenta and separated ovules in l.s., × 25. Outer part of the ovary-wall in l.s. to show the small-celled outer mesocarp limiting the exfoliating exocarp, × 225.

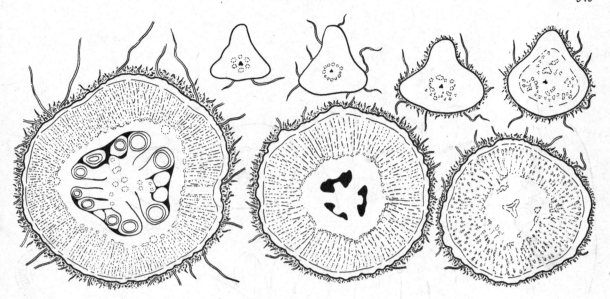

637 *Qualea sp.* Young fruit in t.s. at various levels from the ovular region to the style, the exocarp not yet exfoliated from the vascular mesocarp, ×25.

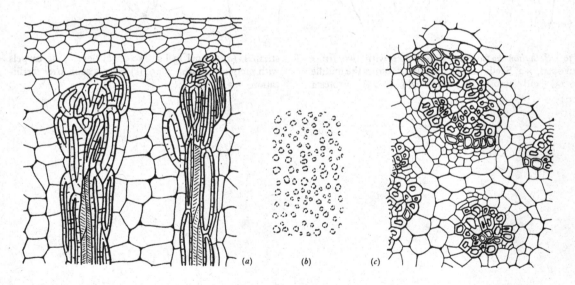

638 *Qualea sp.* Structure of the mesocarp (the exocarp exfoliated): (*a*) fibro-vascular bundles terminating near the small-celled surface, ×225; (*b*) tangential l.s. to show the density of the fibro-vascular bundles, ×25; (*c*) fibro-vascular bundles in t.s., ×225.

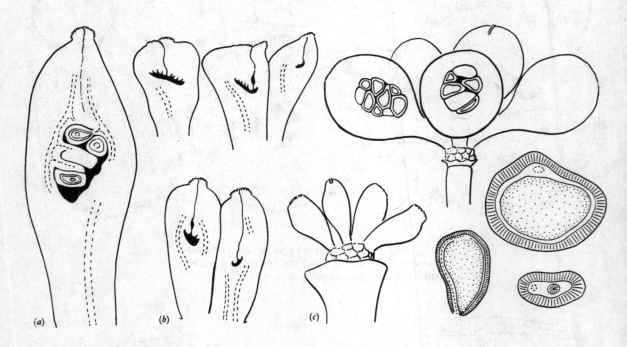

639 *Belliolum haplopus*. Fruit-cluster with two fruits cut open, × 2. Seeds in l.s. (× 5) and t.s. across the middle (× 15) and across the micropyle (× 25); exotesta striated. (*a*) Carpel in l.s., × 25; (*b*) developing carpels with incipient ovules, × 25; (*c*) carpels soon after fertilization, × 5.

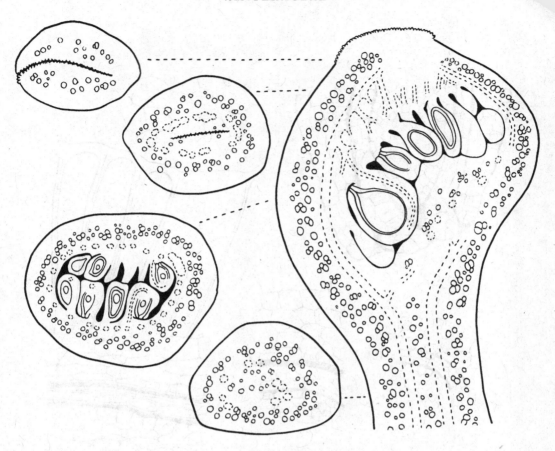

640 *Belliolum haplopus.* Young fruit in l.s. and t.s. at various levels; stone-cells of the cortex indicated as circles, ×25.

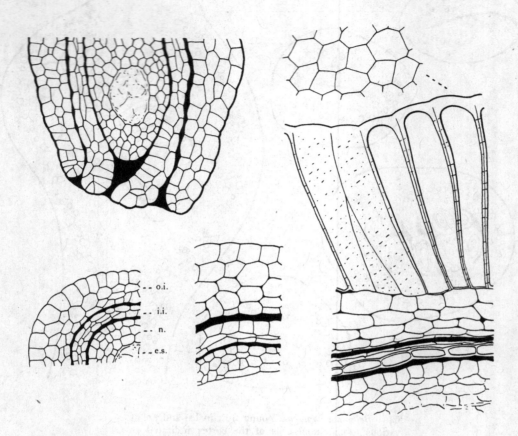

641 *Belliolum haplopus.* Micropyle of the ovule in l.s.;
wall of the ovule, of immature seed, and of mature seed
in t.s. with nucellar remains, ×225; e.s. embyro-sac.

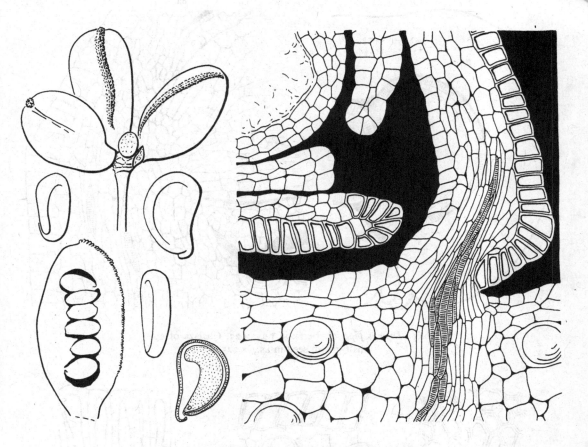

642 *Drimys piperita*. Cluster of berries (2 removed),
× 3. Berry in median l.s., × 5. Seeds, × 10. Funicle and
micropyle of the fully grown but immature seed, × 225.

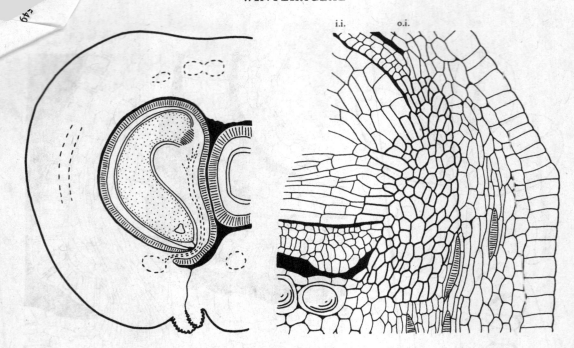

643 *Drimys piperita*. Berry in t.s., ×25. Chalaza of a
half-grown seed in l.s., ×225.

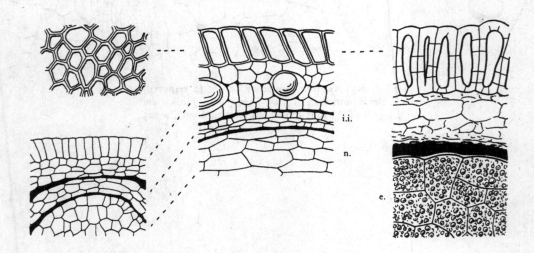

644 *Drimys piperita*. Wall of ovule and seed-coats of an
immature and a mature seed in t.s., ×225.

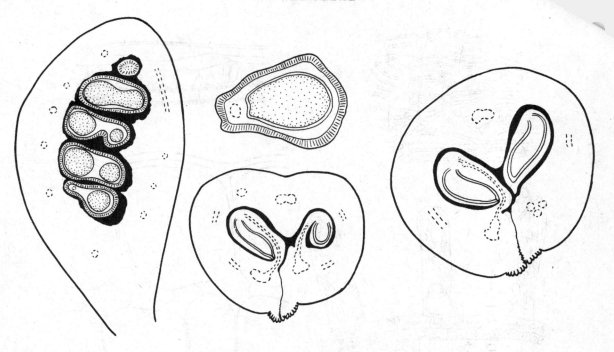

645 *Drimys piperita*. Berry in tangential l.s., × 15. Young berries soon after fertilization, and half-grown berries in t.s., × 25. Seed in t.s., × 25.

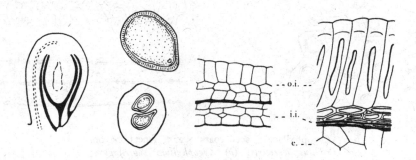

646 *Pseudowintera colorata* (after Bhandari 1963). Ovule in l.s., × 100. Seed in l.s., × 10. Fruit in t.s., with two seeds, × 3. Ovule-wall and seed-coats in t.s., × 225.

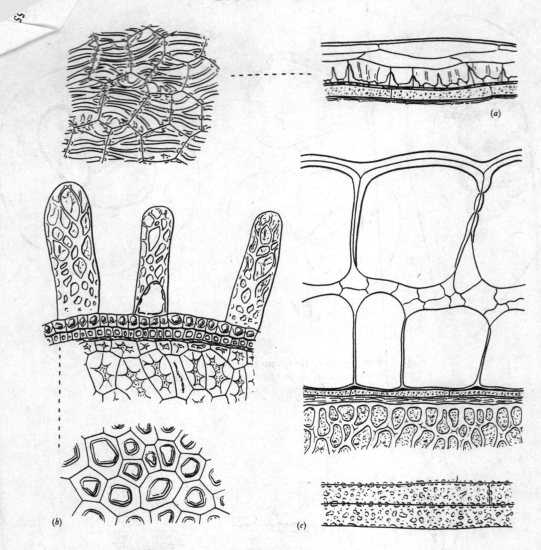

647 Zygophyllaceae, seed-coats ×225; facets, ×500.
(a) *Tribulus terrestris*; (b) *Zygophyllum subtrijugum*;
(c) *Peganum harmala*, with facets of endotegmen.